THE TOOLS OF SCIENCE:
THE SCIENCE PROCESS SKILLS

Test Bank

by

Kevin D. Finson

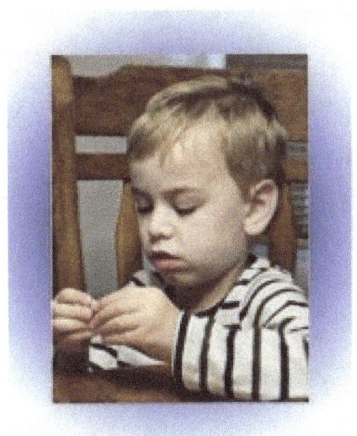

© by Finson, 2024. All rights reserved. No part of this publication may be reproduced or used in any form or by any means, electronic or mechanical, including photocopying, recording, or by any information storage and retrieval system, without the express written permission from the copyright owner. For information regarding permission, write to: kevindfinsonauthor@gmail.com. Users are granted permission to photocopy student tests for use with students, limited to one copy per student, for purposes of teaching science.

Independently Published. ISBN 9798873357932

THE TOOLS OF SCIENCE: THE SCIENCE PROCESS SKILLS

TEST BANK

TABLE OF CONTENTS

Table of Contents (continued)

INTRODUCTION
TO TESTING

TESTING – INTRODUCTION INFORMATION

Students should be assessed on what they have learned in each chapter. There are, of course, different ways to do assessing. The classical way is to provide students end-of-chapter tests. This is part of the strategy the teacher might use for the science process skills. Ideally, the teacher will be assessing students' performance, knowledge, and understandings with each of the individual lessons. Those kinds of assessments are essentially **formative** in nature, and allow the teacher to have a better understanding of what their students are learning and how well they are learning it. The questions included with each lesson lend themselves to formative assessing, as do the products students are creating (tables, graphs, drawings, etc.) with the various lessons. Formative assessing allows the teacher to quickly identify points of weakness in student learning and then take corrective steps to redirect lessons or even repeat them in ways that help students remediate their learning. For some teachers, formative assessment is sufficient.

However, some teachers also like **summative assessing**, which provides them a picture of students' overall and cumulative gains in knowledge and understandings. The bane of schooling for many students is the "final" exam that rears its head at the end of nine-week or semester periods, and even in and even in some cases the year-end exam. More limited in scope are end-of-chapter tests. The tests included in this section are end-of-chapter tests.

The chapter test questions are drawn from two sources. The first is directly from the chapter activities themselves. The second is from background information that the teacher should find ways to share with students at the beginning and throughout each chapter.

The teacher should be aware that each of the chapter tests covers material for all student levels of activities included within the chapter. Hence, a test may include questions about elementary student activities as well as questions about middle school activities. It is inherently unfair to test elementary students on material they have not been taught (i.e. middle school material) or vice versa. Therefore, the teacher using these tests should feel free to omit having students deal with questions about activities or background information that they were not responsible for learning. Consequently, not all questions on a given test may be included in the scoring of the test.

The teacher should also feel free to add questions of his/her own to any of the chapter tests. There may be ancillary material the teacher covers that is not included in any of the science process skill lessons, and it is appropriate to assess student knowledge and understanding of that material.

The teacher will note that the chapter tests are all objective in nature. They are largely multiple choice in format. There are a couple of rare short-answer questions, a rare matching question bank, and questions that require some graphing. A test key is provided for each chapter test. Objective tests, by their nature, take more time to construct than do open-ended question test formats, but they are much quicker to score. A very important practice is for teachers to go through test results with students and explain why incorrect student answers are incorrect and why the keyed answer is the more correct one. This takes some time away from what some view as instruction, but it is actually instruction and reinforces good learning. Students should never be placed in the predicament of not understanding why they failed to answer a question correctly.

Objective tests are only part of the assessment picture. A better-quality assessment program includes performance-based assessments in tandem with objective assessment. Performance-based assessments focus on what students can actually do – how they perform – on specified tasks, and on how well they have learned the concepts and can apply them in different contexts. These kinds of assessments give a depth of insight to student understanding and skill development that is not available through objective testing. For example, it is one thing for a student to select the volume of water in a drawn graduated cylinder but it is an entirely different thing for them to read the water volume in an actual graduated cylinder. The same applies to reading thermometers, finding the masses of objects on balances, and so forth. All this is important. After all, the name "science process skills" actually implies that students engage with and can perform some processes and some skills. Research has clearly shown that most students are more comfortable and confident with performance assessment than objective assessment.

It is beyond the scope of this set of assessments to include performance assessments for each chapter. It is recommended the teacher produce his/her own for use with students. None of those tasks need to be necessarily lengthy or complex, although they could be. Below are some suggestions for the kinds of performance assessments the teacher might generate for each of the chapters. Teachers can be creative in this regard and adjust the complexity for the grade

level of the student.

Chapter 1: Observing

- Provide students with geoblocks or even building blocks and have them identify attributes as well as attribute values of them. Similarly, provide students photographs of different animals and do the same thing – or even go to a zoo to observe live animals.

- Take students to a botanical garden and have them identify the attributes and attribute values of different flowers and plants.

- Give students a set of objects, such as three tree leaves, and give them five minutes to write (or orally share) all the observations they can make about those objects.

- Take students on another blindfolded sense walk to assess growth in students' non-visual observational skills (as compared to the first walk done).

Chapter 2: Inferring

- Have students watch as a glass is filled to almost full with water before bedtime, and then before students come back the next morning pour out some of the water and then have students make various inferences about what happened to the missing water.

- Obtain different kinds of magnets (bars, horseshoes, round donut, etc.) and have students plot their magnetic fields or determine the magnets' strengths by how many paperclips each can hold in a chain formation, etc.

- Take a small object and embed it within a ball of modeling clay, give students a toothpick and graph paper and have them figure out ways to determine the shape of the object without opening the clay ball. The toothpick can be used as a depth probe.

Chapter 3: Measuring

- Have students use metric rulers to Measure the dimensions of many different objects.

- Pre-fill a graduated cylinder and have students read the liquid volume.

- Task students with filling a graduated cylinder to a specific volume with a liquid.

- Task students with determining the mass of an object to within one (or perhaps two) grams of accuracy by using a balance.

- Have students apply natural dyes to filter paper to perform chromatography on them. Natural dyes include juices from vegetables such as beets, smears of chlorophyll from leaves (which usually need alcohol rather than water to work), etc.

Chapter 4: Space-Time Relationships

- Have students calculate the acceleration of different falling objects such as balls or pieces of paper that are of the same size and mass but different in shape (e.g. flat, wadded

into balls, etc.). Note this will likely require students dropping the objects from some height such as from a stepstool, so there should be appropriate precautions put into place before doing this.

- Task students with determining the rate of flow of water through tubes of different diameters. Food coloring in the water is a tremendous help with this, and the tubes should be flexible and at least one meter in length.

- Have students determine if solid objects cool at the same rate as equal masses of water. For example, does a 5-gram metal washer that is heated in hot (not boiling!) water cool at the same rate as 5 grams of water starting at that same temperature?

- Have students measure the transpiration of plant leaves other than from trees.

Chapter 5: Recording and Interpreting Data

- Have students collect discrete data (such as different coin denominations) and make data tables and histograms with those data.

- Have students make observations of something and create a qualitative data table for it.

Chapter 6: Communicating

- Have students read some science articles or reports (such as in children's science magazines, or on the internet) and share in detail what they understand from those articles.

- Have students learn what letters and numbers signal flags have and how to send and interpret signal flag messages.

Chapter 7: Formulating and Using Models

- Have students look up what a Jacob's Staff is and then use it to make a contour map of a small part of the yard.

- Have students select their own area of the yard at which to conduct a transect line study.

- Have students research different simple molecules and create their own models of them.

- Teach students what the symbols are on a real topographic map and then have students interpret what is shown on real topographic maps.

Chapter 8: Using Numbers

- Have students investigate the insulating properties of a variety of materials to see how they affect the melting rate of ice cubes. Materials could include Styrofoam, newspaper, bubble wrap, etc.

- Plot out a "trail" and give students a magnetic compass and have them follow specified paces or distances at compass rose degrees to retrace the "trail." This is

rudimentary orienteering.

- Have students research and compare the sizes of the different moons in the solar system.

Chapter 9: Classifying

- Have students classify marbles.

- Have students classify mineral samples or rock samples.

- Have students classify sea shells.

- Have students classify different dried beans.

- Have students learn how clouds are classified and then make actual observations and identify the clouds using that classification scheme.

- Obtain some classification keys and have students learn to use them and then actually make use of them. There are live-tree finders, winter tree finders, and even animal track finders that can be purchased for a few dollars.

Chapter 10: Identifying and Controlling Variables

- Have students construct bridges using drinking straws and either Styrofoam balls or clay balls, requiring the bridges span a specified distance and then see how much weight they can support before collapsing.

- Have students design and construct containers into which they insert raw eggs and then drop the containers from various heights – with the goal of determining what design can be dropped the farthest while keeping the egg from breaking.

- Have students research and locate different designs for paper airplanes, then construct those airplanes and test them for flight characteristics – including not only distance, but also straight-line flight and also flights that include two or three loop-to-loops.

Chapter 11: Questioning

- Have students read science-related news articles that have some unknown outcomes or for which scientists do not yet know answers, and have students generate scientific questions that could be asked to assist in investigating those things.

- Have students witness some discrepant event science demonstrations and then ask only questions to try to solve the puzzles of how those events work.

- Have students select a science profession they might be interested in and then generate a list of questions they could ask an actual scientist about his/her work – and if possible, locate such a person who is willing to be interviewed by the student. (Don't forget that veterinarians, horticulturalists, etc. are scientists, too.)

Chapter 12: Hypothesizing

- Have students design and conduct some Consumer-Reports types of investigations, such as the whitening power of toothpastes, the cleansing power of dish detergents, the durability of batteries (how long they may last), etc.

- Have students investigate the amount of fading that occurs in different colored construction papers that are exposed to prolonged sunlight.

- Have students investigate the rate of evaporation of water from containers that have the same volume but different sized openings (diameters).

Chapter 13: Predicting

- Have students collect some weather data (such as temperatures, wind direction, barometric pressures, etc.) and predict the next day's weather. Students may need some instruction about those weather variables and how they affect weather.

- Give students a piece of aluminum foil 10 cm x 10 m in size and task them with building a boat or barge from it that can support the most pennies before sinking in a tank of water. The predictions should be about the shapes of the foil that work best, including the area in contact with the water surface, the height of the boat's sides, and even where the pennies are placed in the boat.

- Have students predict how different temperatures affect the stretchiness of rubber bands. Have them use the same size and type of rubber bands and place some in the freezer, some in hot water, etc. and then see how many weights (like metal washers hanging on paperclip hooks) it takes to stretch the rubber bands until they cannot stretch any further.

Chapter 14: Experimenting

- Have students do an independent experiment project where they select a topic to investigate such as one of the 25 listed in the Experimenting Teacher Notes. Be sure include all steps of the experimental process and produce a report to share afterward.

MODULE 1 TEST

THE SCIENCE PROCESS SKILL OF OBSERVING

Module 1 TEST: OBSERVING

NAME: _____

DIRECTIONS: Be sure to read each question carefully. For multiple choice questions, be sure to also carefully ready each of the possible answers. Each is worth 1 point unless otherwise noted. Total Test Value is 17 points.

Short Answer Questions
Directions: Write your answers in the spaces beneath the question.

1. What is an observation? (2 pts)

Multiple Choice Questions
Directions: Write the capital letter for the answer you choose in the blank to the left of the question. Each is worth 1 point.

_____ 2. What is a quantitative observation?
   A. There is no such thing as a quantitative observation
   B. It is an observation that is about an object's qualities
   C. It is an observation that is about measured things of an object
   D. It is an observation about quants.

_____ 3. What is a qualitative observation?
   A. There is no such thing as a quantitative observation
   B. It is an observation that is about an object's qualities
   C. It is an observation that is about measured things of an object
   D. It is an observation about quants.

_____ 4. What are the things that make an inference a good inference?
   A. It is based on good observations
   B. It is based on as many observations as possible
   C. It is made by someone who knows a lot about observations
   D. Both answers A and B
   E. Both answers A and C
   F. Both answers B and C

GO ON TO THE NEXT PAGE!

_____ 5. What are the main attributes of an insect?
    A. It has three body parts and eight legs
    B. It has two body parts and eight legs
    C. It has three body parts and six legs
    D. It has two body parts and six legs

_____ 6. What are the main attributes of a spider?
    A. It has three body parts and eight legs
    B. It has two body parts and eight legs
    C.   It has three body parts and six legs
    D. It has two body parts and six legs

_____ 7. What happens when the north pole of one magnet is put next to the north pole of another magnet?
    A. The two magnets repel each other
    B. The two magnets attract each other
    C. Neither of the two magnets does anything
    D. One magnet demagnetizes the other magnet

_____ 8. What happens to the length of your shadow as the sun climbs higher in the sky?
    A. It gets longer
    B. It stays the same length
    C. It gets shorter

_____ 9. What happens to the shape of your body's sun shadow as the sun's position in the sky moves through the day?
    A.   It gets longer and skinnier and looks less like me
    B.   It gets shorter and thicker and looks more like me
    C. It stays the same each time of the day
    D. It flips over from the right side to the left side

GO ON TO THE NEXT PAGE!

> **Directions:** Use the following situation to answer the next 2 questions.
>
> **You are given a box of new scented markers.**

____10. An attribute of the markers is
   A. color
   B. lemon smell
   C. orange
   D. the box

____11 An attribute value of the markers is
   A. color
   B. lemon smell
   C. size
   D. the box

12. Make a drawing the shows what a convection cell looks like. Draw it in the space below. Be sure to label the parts of the drawing (e.g. "hot" "cold" "rises" "sinks" etc.). (6 pts)

MODULE 1 TEST: OBSERVING

NAME: **KEY**   _____

DIRECTIONS: Be sure to read each question carefully. For multiple choice questions, be sure to also carefully ready each of the possible answers. Each is worth 1 point unless otherwise noted. Total Test Value is 17 points.

Short Answer Questions
Directions: Write your answers in the spaces beneath the question.

1. What is an observation? (2 pts)

   Information gained about an object or event by using one or more of the senses

Multiple Choice Questions
Directions: Write the capital letter for the answer you choose in the blank to the left of the question. Each is worth 1 point.

C _____ 2. What is a quantitative observation?
   A. There is no such thing as a quantitative observation
   B. It is an observation that is about an object's qualities
   C. It is an observation that is about measured things of an object
   D. It is an observation about quants.

B _____ 3. What is a qualitative observation?
   A. There is no such thing as a quantitative observation
   B. It is an observation that is about an object's qualities
   C. It is an observation that is about measured things of an object
   D. It is an observation about quants.

D _____ 4. What are the things that make an inference a good inference?
   A. It is based on good observations
   B. It is based on as many observations as possible
   C. It is made by someone who knows a lot about observations
   D. Both answers A and B
   E. Both answers A and C
   F. Both answers B and C

GO ON TO THE NEXT PAGE!

_C_ 5. What are the main attributes of an insect?
- A. It has three body parts and eight legs
- B. It has two body parts and eight legs
- C. It has three body parts and six legs
- D. It has two body parts and six legs

_B_ 6. What are the main attributes of a spider?
- A. It has three body parts and eight legs
- B. It has two body parts and eight legs
- C. It has three body parts and six legs
- D. It has two body parts and six legs

_A_ 7. What happens when the north pole of one magnet is put next to the north pole of another magnet?
- A. The two magnets repel each other
- B. The two magnets attract each other
- C. Neither of the two magnets does anything
- D. One magnet demagnetizes the other magnet

_C_ 8. What happens to the length of your shadow as the sun climbs higher in the sky?
- A. It gets longer
- B. It stays the same length
- C. It gets shorter

_A_ 9. What happens to the shape of your body's sun shadow as the sun's position in the sky moves through the day?
- A. It gets longer and skinnier and looks less like me
- B. It gets shorter and thicker and looks more like me
- C. It stays the same each time of the day
- D. It flips over from the right side to the left side

GO ON TO THE NEXT PAGE!

> **Directions:** Use the following situation to answer the next 2 questions.
>
> **You are given a box of new scented markers.**

__A__ 10. An attribute of the markers is
   A. color
   B. lemon smell
   C. orange
   D. the box

__B__ 11. An attribute value of the markers is
   A. color
   B. lemon smell
   C. size
   D. the box

12. Make a drawing the shows what a convection cell looks like. Be sure to label the parts of the drawing (e.g. "hot" "cold" "rises" "sinks" etc.). (6 pts)

   The drawing should have two arrows pointing upward at the center, both labeled "Hot", two arrows point downward – one to each side and both labeled "Cold", and two horizontal arrows at the top and two at the bottom where each top arrow is pointing FROM the center and toward the sides and where each bottom arrow is pointing TOWARD the center. The horizontal arrows can be labeled "Current" or left unlabeled. The word "Cell" may be written in the middle of each cell.

MODULE 2 TEST

THE SCIENCE PROCESS SKILL OF INFERRING

Module 2 TEST: INFERRING

Name: _____

DIRECTIONS: Read each question and each of its answers carefully. Select the best answer and write its CAPITAL letter in the blank to the left of the question. Each is worth 1 point. Total test value is 25 points.

Look at the diagram below. It shows a tube filled with three different liquids and a second tube filled with green liquid. Answer the next 5 questions about the diagram.

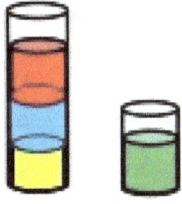

____ 1. The most dense liquid is . . .
    A. Yellow
    B. Blue
    C. Red

____ 2. The least dense liquid is . . .
    A. Yellow
    B. Blue
    C. Red

____ 3. What would happen if the blue liquid is poured into the tube on top of the red liquid?
    A. The blue liquid would stay on top of the red liquid.
    B. The red liquid would sink below the blue liquid.
    C. The blue liquid would sink below the red liquid.
    D. The yellow liquid would rise to replace the blue liquid.

____ 4. What would happen if the red liquid was poured in first, then the yellow liquid, and finally the blue liquid?
    A. Red would stay on bottom, yellow in the middle, and blue on the top.
    B. Red would move to top, yellow move to bottom, and blue move to middle
    C. Red would move to middle, yellow move to top, and blue move to bottom
    D. Red would stay on bottom, blue move to the top, and yellow to the middle

____ 5. If the green liquid is less dense than the red liquid, what will it do if it is poured into the tube with the other colored liquids?
    A. It will float on top of the red liquid.
    B. It will sink until it is below the yellow liquid.
    C. It will settle between the red and blue liquids.
    D. It will settle between the blue and yellow liquids.

GO ON TO THE NEXT PAGE!

___ 6. An inference is a . . .
    A. hypothesis needing to be tested
    B. special kind of observation
    C. single observation made by someone else
    D. conclusion base upon observations

___ 7. The best inference can be made when . . .
    A. observations to make it are ones of good quality
    B. there is only one observation to use to make it
    C. the same person makes it who also made the observation
    D. conclusions are made by other people

___ 8. The best inferences are made when . . .
    A. conclusions are made by other people
    B. there are multiple observations used to make it
    C. the same person who makes them also made the observations
    D. no observations interfere with conclusions to be made

___ 9. One major problem with inferences is that they . . .
    A. are almost always incorrect
    B. are based entirely on observation
    C. can change over time
    D. may or may not be accurate

___ 10. Usually, the best inference is the . . .
    A. one with the most details in it
    B. simplest one
    C. one made by a single person
    D. first one to be made

___ 11. Grandma's television will not turn on. Which of the following statements is an inference about her television?
    A. The television is a black one.
    B. The television is mounted on the wall above Grandma's eye level.
    C. The television is broken.
    D. The television screen is 54 inches across.

___ 12. Which of the following statements it true about a magnet's strength?
    A. It is strongest in the middle of the magnet.
    B. Its strength becomes weaker the farther you are away from it.
    C. It is weakest at the ends or poles of the magnet.
    D. It changes because of the earth's magnetic north pole.

GO ON TO NEXT PAGE

___ 13. The strongest part of a magnet is located at its . . .
     A. poles
     B. middle
     C. edge
     D. center

___ 14. What happens if the south pole of one magnet is put next to the south pole of another magnet?
     A. The magnets pull themselves toward each other.
     B. The magnets push themselves away from each other.
     C. The magnets lose their magnetic strength.
     D. The magnets do not affect each other.

___ 15. Which pole of the needle in a magnetic compass will point to the earth's north magnetic pole?
     A. The needle's south pole.
     B. The needle's north pole.
     C. Its poles cannot point to the earth's north magnetic pole.
     D. Its needle will spin and not point to a pole.

___ 16. If you draw the lines of the magnetic field of a magnet, what will they look like?
     A. Straight lines shooting out from the magnet's ends.
     B. Curved loops going in and out of the magnet's center.
     C. Curved loops going from one end of the magnet to the other end.
     D. Straight lines shooting out from the magnet's center.

___ 17. What happens to the shadow of a post as the sun moves to a position that is more directly overhead?
     A. The shadow gets longer.
     B. The shadow gets darker.
     C. The shadow gets lighter.
     D. The shadow gets shorter.

___ 18. If you mark the position of a shadow through the day and later draw a line connecting those marks, what will the line look like?
     A. It will be a straight line.
     B. It will be a curved line.
     C. It will slant toward one side.
     D. It will be zig-zag (back-and-forth) shape.

GO ON TO THE NEXT PAGE!

> For the next 7 questions, use the following story.
>
> You hear a crash in the next room. You get up and begin to dash to the room. On your way, your cat scampers past you out of that room. When you enter the room, you see a glass vase on the floor beneath a shelf.

Read each of the following statements. For those that are **Observations**, write an "O" in the blank to the left of the statement. For those that are **Inferences,** write an "I" in the blank to the left of the statement. Each is worth 1 point.

____ 19. The vase is in shatters on the floor.

____ 20. A small earthquake made the vase fall off the shelf.

____ 21. The shelf is 4 inches wide and the vase was 6 inches wide.

____ 22. The cat knocked the vase off the shelf.

____ 23. The shelf is high up on the wall.

____ 24. The vase was not put on the shelf properly.

____ 25. A pottery vase would not have shattered like the glass one did.

Module 2 TEST: INFERRING
Name: _____**KEY**_____

> **DIRECTIONS:** Read each question and each of its answers carefully. Select the best answer and write its CAPITAL letter in the blank to the left of the question. Each is worth 1 point. Total test value is 25 points.

Look at the diagram below. It shows a tube filled with three different liquids and a second tube filled with green liquid. Answer the next 5 questions about the diagram.

__A__ 1. The most dense liquid is . . .
   A. Yellow
   B. Blue
   C. Red

__C__ 2. The least dense liquid is . . .
   A. Yellow
   B. Blue
   C. Red

__C__ 3. What would happen if the blue liquid is poured into the tube on top of the red liquid?
   A. The blue liquid would stay on top of the red liquid.
   B. The red liquid would sink below the blue liquid.
   C. The blue liquid would sink below the red liquid.
   D. The yellow liquid would rise to replace the blue liquid.

__B__ 4. What would happen if the red liquid was poured in first, then the yellow liquid, and finally the blue liquid?
   A. Red would stay on bottom, yellow in the middle, and blue on the top.
   B. Red would move to top, yellow move to bottom, and blue move to middle
   C. Red would move to middle, yellow move to top, and blue move to bottom
   D. Red would stay on bottom, blue move to the top, and yellow to the middle

__A__ 5. If the green liquid is less dense than the red liquid, what will it do if it is poured into the tube with the other colored liquids?
   A. It will float on top of the red liquid.
   B. It will sink until it is below the yellow liquid.
   C. It will settle between the red and blue liquids.
   D. It will settle between the blue and yellow liquids.

GO ON TO THE NEXT PAGE!

_D_ 6. An inference is a . . .
  A. hypothesis needing to be tested
  B. special kind of observation
  C. single observation made by someone else
  D. conclusion base upon observations

_A_ 7. The best inference can be made when . . .
  A. observations to make it are ones of good quality
  B. there is only one observation to use to make it
  C. the same person makes it who also made the observation
  D. conclusions are made by other people

_B_ 8. The best inferences are made when . . .
  A. conclusions are made by other people
  B. there are multiple observations used to make it
  C. the same person who makes them also made the observations
  D. no observations interfere with conclusions to be made

_D_ 9. One major problem with inferences is that they . . .
  A. are almost always incorrect
  B. are based entirely on observation
  C. can change over time
  D. may or may not be accurate

_B_ 10. Usually, the best inference is the . . .
  A. one with the most details in it
  B. simplest one
  C. one made by a single person
  D. first one to be made

_C_ 11. Grandma's television will not turn on. Which of the following statements is an inference about her television?
  A. The television is a black one.
  B. The television is mounted on the wall above Grandma's eye level.
  C. The television is broken.
  D. The television screen is 54 inches across.

_B_ 12. Which of the following statements it true about a magnet's strength?
  A. It is strongest in the middle of the magnet.
  B. Its strength becomes weaker the farther you are away from it.
  C. It is weakest at the ends or poles of the magnet.
  D. It changes because of the earth's magnetic north pole.

GO ON TO THE NEXT PAGE!

_A_ 13. The strongest part of a magnet is located at its . . .
   A. poles
   B. middle
   C. edge
   D. center

_B_ 14. What happens if the south pole of one magnet is put next to the south pole of another magnet?
   A. The magnets pull themselves toward each other.
   B. The magnets push themselves away from each other.
   C. The magnets lose their magnetic strength.
   D. The magnets do not affect each other.

_A_ 15. Which pole of the needle in a magnetic compass will point to the earth's north magnetic pole?
   A. The needle's south pole.
   B. The needle's north pole.
   C. Its poles cannot point to the earth's north magnetic pole.
   D. Its needle will spin and not point to a pole.

_C_ 16. If you draw the lines of the magnetic field of a magnet, what will they look like?
   A. Straight lines shooting out from the magnet's ends.
   B. Curved loops going in and out of the magnet's center.
   C. Curved loops going from one end of the magnet to the other end.
   D. Straight lines shooting out from the magnet's center.

_D_ 17. What happens to the shadow of a post as the sun moves to a position that is more directly overhead?
   A. The shadow gets longer.
   B. The shadow gets darker.
   C. The shadow gets lighter.
   D. The shadow gets shorter.

_B_ 18. If you mark the position of a shadow through the day and later draw a line connecting those marks, what will the line look like?
   A. It will be a straight line.
   B. It will be a curved line.
   C. It will slant toward one side.
   D. It will be zig-zag (back-and-forth) shape.

GO ON TO THE NEXT PAGE!

> For the next 7 questions, use the following story.
>
> You hear a crash in the next room. You get up and begin to dash to the room. On your way, your cat scampers past you out of that room. When you enter the room, you see a glass vase on the floor beneath a shelf.

Read each of the following statements. For those that are **Observations**, write an "O" in the blank to the left of the statement. For those that are **Inferences,** write an "I" in the blank to the left of the statement. Each is worth 1 point.

_O_ 19. The vase is in shatters on the floor.

_I_ 20. A small earthquake made the vase fall off the shelf.

_O_ 21. The shelf is 4 inches wide and the vase was 6 inches wide.

_I_ 22. The cat knocked the vase off the shelf.

_O_ 23. The shelf is high up on the wall.

_I_ 24. The vase was not put on the shelf properly.

_I_ 25. A pottery vase would not have shattered like the glass one did.

MODULE 3 TEST

THE SCIENCE PROCESS SKILL OF MEASURING

Module 3 TEST: MEASURING

Name: _____

> **DIRECTIONS:** Read each question carefully, and read each answer carefully. Select the best answer and write its CAPITAL letter in the blank to the left of the question. Each question is worth 1 point unless otherwise stated. Total test value is 32 points.

____ 1. An astronomical unit (AU) is the . . .
     A. width of our solar system
     B. width of the sun
     C. distance from the sun to the earth
     D. average distance between outer planets

____ 2. The inner planets are . . .
     A. Mercury, Venus, Earth, and Mars
     B. Jupiter, Saturn, Uranus, Neptune, and Pluto
     C. Earth, Mars, Jupiter, and Saturn
     D. Mercury and Venus

____ 3. The outer planets are . . .
     A. Mercury, Venus, Earth, and Mars
     B. Jupiter, Saturn, Uranus, Neptune, and Pluto
     C. Earth, Mars, Jupiter, and Saturn
     D. Mercury and Venus

____ 4. The largest planet in our solar system is . . .
     A. Neptune
     B. Uranus
     C. Saturn
     D. Jupiter

____ 5. The two smallest planets in our solar system are . . .
     A. Mercury and Venus
     B. Mars and Earth
     C. Pluto and Mars
     D. Mercury and Pluto

____ 6. How do the distances between the inner planets compare to the distances between the outer planets?
     A. The distances between inner planets are smaller than between outer planets.
     B. The distances between inner planets are larger than between the outer planets.
     C. The distances between all the planets is about the same.
     D. The distances between planets increase as you go from Mercury to Pluto.

GO ON TO THE NEXT PAGE!

_____7. Someone gives you a leaf and asks you to measure the length of its outside edge. What might be the best way to do that measurement?
  A. Put a piece of graph paper on top of it and count the squares on the paper.
  B. Use a ruler to see how long and wide the leaf is.
  C. Put a string around the outside edge of the leaf and then measure the string.
  D. Set the leaf beside another leaf whose length you already know.

_____8. You have a table that is 10 pencils long. One pencil is 6 paperclips long. How many paperclips long is the table?
  A. 60
  B. 16
  C.   6
  D. 10

_____9. What could happen if you used paperclips to measure the length of a room but someone else used hair combs to measure its length?
  A. Everyone would be confused about the room length.
  B. You would need to find out how many paperclip lengths are in one hair comb.
  C. You would need to find out how many hair comb lengths are in one paperclip.
  D. The room measurement would be in different units of length.

_____10. What kind of measurement are you making when you put an object on one side of a balance and a bunch of metal washers on the other side of the balance?
  A. length
  B. weight
  C. volume
  D. time

_____11. What kind of measurement are you making when you fill a jar with marbles?
  A. length
  B. weight
  C. volume
  D. time

_____12. The technique that can be used to separate the parts of a substance is called . . .
  A. fibonacci
  B. septa separation
  C. chromatography
  D. filteration

GO ON TO THE NEXT PAGE!

Use the following diagram at the right to answer the next 5 questions. The diagram shows a filter paper strip that had a black dot drawn on it and then its end placed into water.

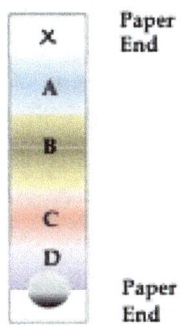

_____ 13. Which of the colors dissolved most easily with the water?
A. The color at position A
B. The color at position B
C. The color at position C
D. The color at position D

_____ 14. Which of the colors was the first to appear from the black dot?
A. The color at position A
B. The color at position B
C. The color at position C
D. The color at position D

_____ 15. Suppose you had a second filter paper strip that had the yellow color at position A on the filter paper. What could you say about the yellow on the first strip and the yellow on the second strip?
A. The water got to the yellow on the first strip earlier than on the second strip.
B. They are probably different kinds of yellows.
C. The black inks that made the dots were different.
D. The second yellow had to first move through the blue color.

_____ 16. Which of the colors on the filter paper strip moved the quickest up the strip?
A. The color at position A
B. The color at position B
C. The color at position C
D. The color at position D

_____ 17. If there was a color in the black dot that would not dissolve in water, where would it be on the filter paper strip?
A. At the top of position A
B. Where position B and C meet
C. At the top of position D
D. In the black dot at the bottom of the strip

GO ON TO THE NEXT PAGE!

____18. A unit of measurement that is uniform in size and accepted everywhere is called a . . .
   A. uniform unit
   B. measured unit
   C. nonstandard unit
   D. standardized unit

____19. The SI system of measurement is also known as the . . .
   A. metric system
   B. English system
   C. Imperial system
   D. international system

____20. The base units in the SI system are the . . .
   A. foot, gallon, and pound
   B. inch, pint, and ounce
   C. meter, liter, and gram
   D. kilometer, second, and slug

____21. Temperature in the SI system is measured in the scale called . . .
   A. Celsius
   B. Fahrenheit
   C. Deltagrade
   D. Graduscale

____22. The curved surface of a liquid like water in a tube like a graduated cylinder is called a(an) . . .
   A. curl
   B. adhesion
   C. gradation
   D. meniscus

____23. The amount of matter or material in an object is known as . . .
   A. volume
   B. mass
   C. weight
   D. slug

____24. The gravitational pull or force on an object is known as . . .
   A. volume
   B. mass
   C. weight
   D. slug

GO ON TO THE NEXT PAGE!

____25. A unit of measurement that is a combination of two other kinds of units is called a . . .
    A. combination unit
    B. split unit
    C. errored unit
    D. derived unit

____26. A unit such as centimeters per second (cm/s) or miles per hour (mi/hr) is a . . .
    A. combination unit
    B. split unit
    C. errored unit
    D. derived unit

MATCHING: Measurement Prefixes

DIRECTIONS: Match the items in Column A with the items in Column B. Write the letter of the item in Column B in the blank that it matches in Column A.

Column A (unit prefix)	Column B (unit size)
____27. kilo-	A. 10 (ten)
	B. 100 (one hundred)
____28. milli-	C. 1,000 (one thousand)
	D. 1,000,000 (one million)
____29. mega –	E. 1/10 (one tenth)
	F. 1/100 (one one-hundredth)
____30. deka-	G. 1/1,000 (one one-thousandth)
____31. centi-	
____32. deci-	

Module 3 TEST: MEASURING

Name: _____KEY_____

DIRECTIONS: Read each question carefully, and read each answer carefully. Select the best answer and write its CAPITAL letter in the blank to the left of the question. Each question is worth 1 point unless otherwise stated.   Total test value is 32 points.

__C__ 1. An astronomical unit (AU) is the . . .
   A. width of our solar system
   B. width of the sun
   C. distance from the sun to the earth
   D. average distance between outer planets

__A__ 2. The inner planets are . . .
   A. Mercury, Venus, Earth, and Mars
   B. Jupiter, Saturn, Uranus, Neptune, and Pluto
   C. Earth, Mars, Jupiter, and Saturn
   D. Mercury and Venus

__B__ 3. The outer planets are . . .
   A. Mercury, Venus, Earth, and Mars
   B. Jupiter, Saturn, Uranus, Neptune, and Pluto
   C. Earth, Mars, Jupiter, and Saturn
   D. Mercury and Venus

__D__ 4. The largest planet in our solar system is . . .
   A. Neptune
   B. Uranus
   C. Saturn
   D. Jupiter

__D__ 5. The two smallest planets in our solar system are . . .
   A. Mercury and Venus
   B. Mars and Earth
   C. Pluto and Mars
   D. Mercury and Pluto

__A__ 6. How do the distances between the inner planets compare to the distances between the outer planets?
   A. The distances between inner planets are smaller than between outer planets.
   B. The distances between inner planets are larger than between the outer planets.
   C. The distances between all the planets is about the same.
   D. The distances between planets increase as you go from Mercury to Pluto.

GO ON TO THE NEXT PAGE!

C 7. Someone gives you a leaf and asks you to measure the length of its outside edge. What might be the best way to do that measurement?
   A. Put a piece of graph paper on top of it and count the squares on the paper.
   B. Use a ruler to see how long and wide the leaf is.
   C. Put a string around the outside edge of the leaf and then measure the string.
   D. Set the leaf beside another leaf whose length you already know.

A 8. You have a table that is 10 pencils long. One pencil is 6 paperclips long. How many paperclips long is the table?
   A. 60
   B. 16
   C.   6
   D. 10

D 9. What could happen if you used paperclips to measure the length of a room but someone else used hair combs to measure its length?
   A. Everyone would be confused about the room length.
   B. You would need to find out how many paperclip lengths are in one hair comb.
   C. You would need to find out how many hair comb lengths are in one paperclip.
   D. The room measurement would be in different units of length.

B 10. What kind of measurement are you making when you put an object on one side of a balance and a bunch of metal washers on the other side of the balance?
   A. length
   B. weight
   C. volume
   D. time

C 11. What kind of measurement are you making when you fill a jar with marbles?
   A. length
   B. weight
   C. volume
   D. time

C 12. The technique that can be used to separate the parts of a substance is called . . .
   A. fibonacci
   B. septa separation
   C. chromatography
   D. filteration

GO ON TO THE NEXT PAGE!

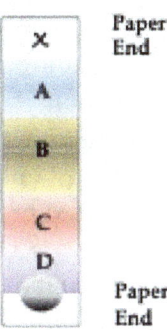

Use the following diagram at the right to answer the next 5 questions. The diagram shows a filter paper strip that had a black dot drawn on it and then its end placed into water.

_A_ 13. Which of the colors dissolved most easily with the water?
    A. The color at position A
    B. The color at position B
    C. The color at position C
    D. The color at position D

_A_ 14. Which of the colors was the first to appear from the black dot?
    A. The color at position A
    B. The color at position B
    C. The color at position C
    D. The color at position D

_B_ 15. Suppose you had a second filter paper strip that had the yellow color at position A on the filter paper. What could you say about the yellow on the first strip and the yellow on the second strip?
    A. The water got to the yellow on the first strip earlier than on the second strip.
    B. They are probably different kinds of yellows.
    C. The black inks that made the dots were different.
    D. The second yellow had to first move through the blue color.

_A_ 16. Which of the colors on the filter paper strip moved the quickest up the strip?
    A. The color at position A
    B. The color at position B
    C. The color at position C
    D. The color at position D

_D_ 17. If there was a color in the black dot that would not dissolve in water, where would it be on the filter paper strip?
    A. At the top of position A
    B. Where position B and C meet
    C. At the top of position D
    D. In the black dot at the bottom of the strip

GO ON TO THE NEXT PAGE!

D 18. A unit of measurement that is uniform in size and accepted everywhere is called a . . .
     A. uniform unit
     B. measured unit
     C. nonstandard unit
     D. standardized unit

A 19. The SI system of measurement is also known as the . . .
     A. metric system
     B. English system
     C. Imperial system
     D. international system

C 20. The base units in the SI system are the . . .
     A. foot, gallon, and pound
     B. inch, pint, and ounce
     C. meter, liter, and gram
     D. kilometer, second, and slug

A 21. Temperature in the SI system is measured in the scale called . . .
     A. Celsius
     B. Fahrenheit
     C. Deltagrade
     D. Graduscale

D 22. The curved surface of a liquid like water in a tube like a graduated cylinder is called a(an) . . .
     A. curl
     B. adhesion
     C. gradation
     D. meniscus

B 23. The amount of matter or material in an object is known as . . .
     A. volume
     B. mass
     C. weight
     D. slug

C 24. The gravitational pull or force on an object is known as . . .
     A. volume
     B. mass
     C. weight
     D. slug

GO ON TO THE NEXT PAGE!

_D_ 25. A unit of measurement that is a combination of two other kinds of units is called a . . .
    A. combination unit
    B. split unit
    C. errored unit
    D. derived unit

_D_ 26. A unit such as centimeters per second (cm/s) or miles per hour (mi/hr) is a . . .
    A. combination unit
    B. split unit
    C. errored unit
    D. derived unit

MATCHING: Measurement Prefixes

DIRECTIONS: Match the items in Column A with the items in Column B.
Write the letter of the item in Column B in the blank that it
matches in Column A.

Column A
(unit prefix)

Column B
(unit size)

_C_ 27. kilo-

_G_ 28. milli-

_D_ 29. mega –

_A_ 30. deka-

_F_ 31. centi-

_E_ 32. deci-

A. 10 (ten)
B. 100 (one hundred)
C. 1,000 (one thousand)
D. 1,000,000 (one million)
E. 1/10 (one tenth)
F. 1/100 (one one-hundredth)
G. 1/1,000 (one one-thousandth)

MODULE 4 TEST

THE SCIENCE PROCESS SKILL OF RECORDING AND INTERPRETING DATA

Module 4 TEST: RECORDING AND INTERPRETING DATA

NAME: _____

> **DIRECTIONS:** Read each question carefully. Read each answer carefully. Select the best answer and write its CAPITAL letter in the blank to the left of the question. Each question is worth 1 point unless otherwise noted. Total test value is 58 points.

_____ 1. Where are most stars located on a Hertzsprug-Russel diagram?
  A. in the lower left corner
  B. in the upper right corner
  C. in the middle on the Main Sequence
  D. in the left center

_____ 2. Where are blue supergiant stars located on a Hertzsprug-Russel diagram?
  A. in the upper left corner
  B. in the lower left corner
  C. in the upper right corner
  D. in the middle on the Main Sequence

_____ 3. Where are red supergiant stars located on a Hertzsprug-Russel diagram?
  A. in the upper left corner
  B. in the upper right corner
  C. in the lower right corner
  D. in the middle on the Main Sequence

_____ 4. What is the spectral class of our sun?
  A. M
  B. O
  C. A
  D. G

_____ 5. What is one way to determine the strength of a magnet?
  A. See how far away it can attract a paperclip
  B. See how many other magnets it can hold to itself
  C. Hang a magnet on a paperclip attached to another magnet
  D. Find out how heavy it is

_____ 6. What happens to the strength of magnets when you put two of them together?
  A. Their combined strengths are the same as one magnet's strength.
  B. Their combined strengths is less than twice one magnet's strength.
  C. Their combined strengths are twice one magnet's strength.
  D. Their combined strengths are less than one magnet's strength.

GO ON TO THE NEXT PAGE!

> **Use the following data table to answer the next 5 questions.**

Table 1

Planet Facts

Planet Name	Diameter of the Planet	Number of Moons	Length of One Day (in Earth Time)	Planet Distance from Sun (in AU)
Mercury	4,880 km	0	59 days	0.4
Venus	12,104 km	0	243 days	0.7
Earth	12,712 km	1	23 hours + 56 minutes	1
Mars	6,794 km	2	24 hours + 37 minutes	1.5
Jupiter	142,800 km	80	9 hours + 55 minutes	5.0
Saturn	120,536 km	146	10 hours + 14 minutes	9.5
Uranus	51,118 km	27	24 hours	19.0
Neptune	49,244 km	14	22 hours	30.0
Pluto	4,740 km	5	6 days + 9 hours	40.0

_____ 7. How many planets are smaller than Earth?
     A. 3
     B. 4
     C. 5
     D. 6

_____ 8. How many planets are larger than Earth?
     A. 3
     B. 4
     C. 5
     D. 6

_____ 9. Which planet is closest in size to Earth's size?
     A. Pluto
     B. Mars
     C. Venus
     D. Mercury

GO ON TO THE NEXT PAGE!

_____ 10. Which planets have the fewest number of moons?
    A. Mercury and Pluto
    B. Venus and Pluto
    C. Mercury and Mars
    D. Mercury and Venus

_____ 11. Which planet has the most moons?
    A. Jupiter
    B. Saturn
    C. Uranus
    D. Neptune

_____ 12. The amount of time it takes for half of the radioactivity in a substance to become non-radioactive is called . . .
    A. half-life
    B. decay rate
    C. radiosonde
    D. depletion rate

_____ 13. When a radioactive atom becomes non-radioactive, the process is called . . .
    A. radiosonde
    B. depletion rate
    C. half-life
    D. decay

_____ 14. How much radioactivity will go away and become non-radioactive in two half-lives?
    A. one-fourth (or 25%)
    B. one-half (or 50%)
    C. three-fourths (or 75%)
    D. all of it (or 100%)

_____ 15. The highest ocean tide is called a . . .
    A. neap tide
    B. lunar tide
    C. spring tide
    D. solar tide

_____ 16. The lowest ocean tide is called a . . .
    A. neap tide
    B. lunar tide
    C. spring tide
    D. solar tide

GO ON TO THE NEXT PAGE!

_____ 17. When the moon in the new moon phase, what kind of tide to we have on the earth?
     A. neap tide
     B. lunar tide
     C. spring tide
     D. solar tide

_____ 18. When the moon is in either the first quarter or third quarter phase, what kind of tide do we have on the earth?
     A. neap tide
     B. lunar tide
     C. spring tide
     D. solar tide

_____ 19. How are the sun, the earth, and the moon aligned when we have a spring tide on the earth?
     A. They are lined up at right angles to each other.
     B. They are all lined up with the sun between the earth and the moon.
     C. They are all lined up in a row with the earth between the sun and the moon.
     D. They are all lined up in a row with the moon between the sun and the earth.

_____ 20. A data table that includes mostly number values is called a . . .
     A. descriptive table
     B. quantitative table
     C. number table
     D. qualitative table

_____ 21. A data table that includes mostly descriptive information is called a . . .
     A. descriptive table
     B. quantitative table
     C. number table
     D. qualitative table

_____ 22. What is the name of the process when you try to determine the value of a variable between two data points that are plotted on the graph?
     A. interpolation
     B. manipulation
     C. extrapolation
     D. linuation

GO ON TO THE NEXT PAGE!

_____ 23. What is the name of the process when you try to determine the value of a variable that is somewhere beyond any of the data points plotted on a graph?
   A. interpolation
   B. manipulation
   C. extrapolation
   D. linuation

24. Make a data table for the following information. Make the data table in the space below. This is worth 12 points.

   A scientist collected data about the growth of a plant. Plant growth was measured by its height. Data were collected each day for six days. The first day the plant was 2 cm tall. The second day the plant was 4 cm tall. The third day the plant was 6 cm tall. The fourth day the plant was 8 cm tall. The fifth day the plant was 9 cm tall. The sixth day the plant was 11 cm tall.

GO ON TO THE NEXT PAGE!

25. Use the following data table and the graph grid to make a graph. This is worth 11 points.

Table 1 Colors of Toy Balls	
Color	**Number of Balls**
Red	2
Orange	5
Yellow	4
Green	3
Blue	6
Purple	1

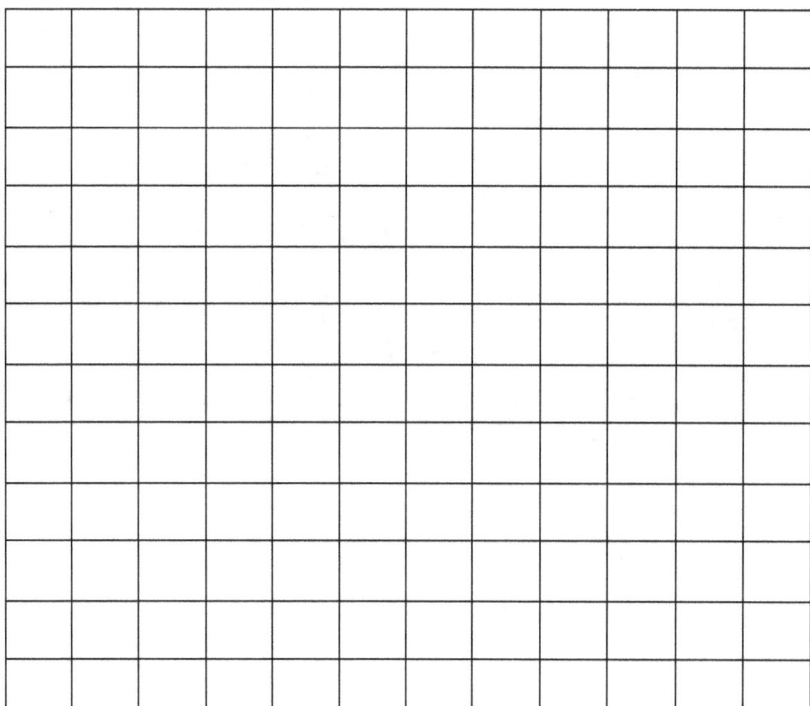

GO ON TO THE NEXT PAGE!

26. Use the following data table and the graph grid to make a graph. This is worth 12 points.

Table 2 Hamster Weight Over Time	
Time (in days)	**Weight (in ounces)**
1	3
2	4
2	5
4	7
5	9
6	10
7	10
8	12
9	13
10	13

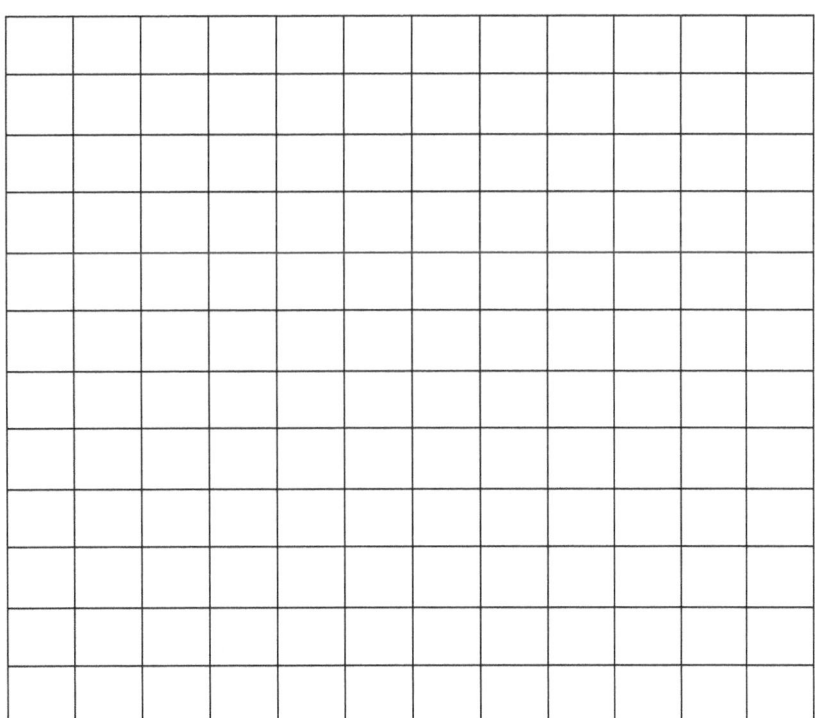

Module 4 TEST: RECORDING AND INTERPRETING DATA

NAME: KEY

> **DIRECTIONS:** Read each question carefully. Read each answer carefully. Select the best answer and write its CAPITAL letter in the blank to the left of the question. Each question is worth 1 point unless otherwise noted. Total test value is 58 points.

C 1. Where are most stars located on a Hertzsprug-Russel diagram?
   A. in the lower left corner
   B. in the upper right corner
   C. in the middle on the Main Sequence
   D. in the left center

A 2. Where are blue supergiant stars located on a Hertzsprug-Russel diagram?
   A. in the upper left corner
   B. in the lower left corner
   C. in the upper right corner
   D. in the middle on the Main Sequence

B 3. Where are red supergiant stars located on a Hertzsprug-Russel diagram?
   A. in the upper left corner
   B. in the upper right corner
   C. in the lower right corner
   D. in the middle on the Main Sequence

D 4. What is the spectral class of our sun?
   A. M
   B. O
   C. A
   D. G

A 5. What is one way to determine the strength of a magnet?
   A. See how far away it can attract a paperclip
   B. See how many other magnets it can hold to itself
   C. Hang a magnet on a paperclip attached to another magnet
   D. Find out how heavy it is

B 6. What happens to the strength of magnets when you put two of them together?
   A. Their combined strengths are the same as one magnet's strength.
   B. Their combined strengths is less than twice one magnet's strength.
   C. Their combined strengths are twice one magnet's strength.
   D. Their combined strengths are less than one magnet's strength.

GO ON TO THE NEXT PAGE!

Use the following data table to answer the next 5 questions.

Table 1
Planet Facts

Planet Name	Diameter of the Planet	Number of Moons	Length of One Day (in Earth Time)	Planet Distance from Sun (in AU)
Mercury	4,880 km	0	59 days	0.4
Venus	12,104 km	0	243 days	0.7
Earth	12,712 km	1	23 hours + 56 minutes	1
Mars	6,794 km	2	24 hours + 37 minutes	1.5
Jupiter	142,800 km	80	9 hours + 55 minutes	5.0
Saturn	120,536 km	146	10 hours + 14 minutes	9.5
Uranus	51,118 km	27	24 hours	19.0
Neptune	49,244 km	14	22 hours	30.0
Pluto	4,740 km	5	6 days + 9 hours	40.0

B   7. How many planets are smaller than Earth?
- A. 3
- B. 4
- C. 5
- D. 6

B   8. How many planets are larger than Earth?
- A. 3
- B. 4
- C. 5
- D. 6

C   9. Which planet is closest in size to Earth's size?
- A. Pluto
- B. Mars
- C. Venus
- D. Mercury

GO ON TO THE NEXT PAGE!

D 10. Which planets have the fewest number of moons?
    A. Mercury and Pluto
    B. Venus and Pluto
    C. Mercury and Mars
    D. Mercury and Venus

B 11. Which planet has the most moons?
    A. Jupiter
    B. Saturn
    C. Uranus
    D. Neptune

A 12. The amount of time it takes for half of the radioactivity in a substance to become non-radioactive is called . . .
    A. half-life
    B. decay rate
    C. radiosonde
    D. depletion rate

D 13. When a radioactive atom becomes non-radioactive, the process is called . . .
    A. radiosonde
    B. depletion rate
    C. half-life
    D. decay

B 14. How much radioactivity will go away and become non-radioactive in two half-lives?
    A. one-fourth (or 25%)
    B. one-half (or 50%)
    C. three-fourths (or 75%)
    D. all of it (or 100%)

C 15. The highest ocean tide is called a . . .
    A. neap tide
    B. lunar tide
    C. spring tide
    D. solar tide

A 16. The lowest ocean tide is called a . . .
    A. neap tide
    B. lunar tide
    C. spring tide
    D. solar tide

GO ON TO THE NEXT PAGE!

C   17. When the moon in the new moon phase, what kind of tide to we have on the earth?
    A. neap tide
    B. lunar tide
    C. spring tide
    D. solar tide

A   18. When the moon is in either the first quarter or third quarter phase, what kind of tide do we have
    on the earth?
    A. neap tide
    B. lunar tide
    C. spring tide
    D. solar tide

D   19. How are the sun, the earth, and the moon aligned when we have a spring tide on the earth?
    A. They are lined up at right angles to each other.
    B. They are all lined up with the sun between the earth and the moon.
    C. They are all lined up in a row with the earth between the sun and the moon.
    D. They are all lined up in a row with the moon between the sun and the earth.

B   20. A data table that includes mostly number values is called a . . .
    A. descriptive table
    B. quantitative table
    C. number table
    D. qualitative table

D   21. A data table that includes mostly descriptive information is called a . . .
    A. descriptive table
    B. quantitative table
    C. number table
    D. qualitative table

A   22. What is the name of the process when you try to determine the value of a variable between two
    data points that are plotted on the graph?
    A. interpolation
    B. manipulation
    C. extrapolation
    D. linuation

GO ON TO THE NEXT PAGE!

_**C**_ 23. What is the name of the process when you try to determine the value of a variable that is somewhere beyond any of the data points plotted on a graph?
   A. interpolation
   B. manipulation
   C. extrapolation
   D. linuation

24. Make a data table for the following information. Make the data table in the space below. This is worth 12 points.

   A scientist collected data about the growth of a plant. Plant growth was measured by its height. Data were collected each day for six days. The first day the plant was 2 cm tall. The second day the plant was 4 cm tall. The third day the plant was 6 cm tall. The fourth day the plant was 8 cm tall. The fifth day the plant was 9 cm tall. The sixth day the plant was 11 cm tall.

   Score one point for each of the following elements of the table:
   • title
   • two columns
   • heading at the top of the first column
   • first column heading shows variable name (Time or Days)
   • first column heading shows units (Day #)
   • heading at the top of the second column (Plant Growth)
   • second column heading shows variable name (Plant Height)
   • second column heading shows units (cm)
   • the independent variable is in the first column (time or days)
   • the values in the first column are in either ascending or descending order (1, 2, 3, 4, 5, 6)
   • the dependent variable is in the second column (height of plant)
   • The values for the second column are matched to their corresponding values in the first column

Table 1 Plant Growth	
Time (in days)	Growth (in cm)
1	2
2	4
3	6
4	8
5	9
6	11

25. Use the following data table and the graph grid to make a graph. This is worth 11 points.

Table 1 Colors of Toy Balls	
Color	**Number of Balls**
Red	2
Orange	5
Yellow	4
Green	3
Blue	6
Purple	1

Score one point for each of the following elements of the graph:
• title
• independent variable is on horizontal axis (color)
• horizonal axis is labeled
• horizonal axis label gives name of the variable (color of balls)
• horizonal axis label gives units of measurement of variable (no units on this)
• dependent variable is on vertical axis (number)
• vertical axis is labeled (number of balls)
• vertical axis label gives name of the variable (only numbers 1, 2, 3, 4, 5, 6)
• vertical axis label gives units of measurement of variable (no units on this)
• bars are drawn for each ball color
• bars are not connected with a line

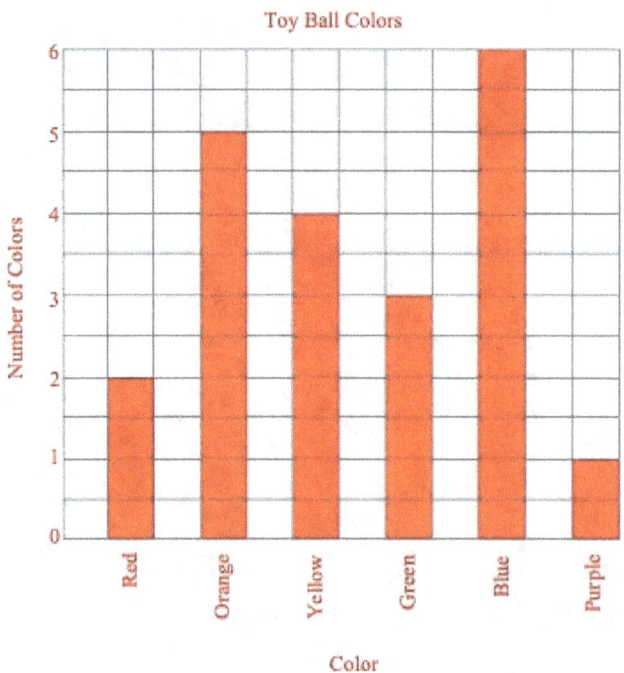

GO ON TO THE NEXT PAGE!

26. Use the following data table and the graph grid to make a graph. This is worth 12 points.

Table 2 Hamster Weight Over Time	
Time (in days)	**Weight (in ounces)**
1	3
2	4
2	5
4	7
5	9
6	10
7	10
8	12
9	13
10	13

Score one point for each of the following graph elements:
• title
• independent variable is on horizontal axis (time)
• horizontal axis label gives name of the variable (Time)
• dependent variable is on vertical axis (weight)
• vertical axis label gives name of the variable (weight)
• vertical axis label gives units of measurement
    of variable (ounces)

• horizontal axis is labeled (Time)
• horizontal axis label gives units of
    measurement of variable (# Days)
• vertical axis is labeled (Hamster Weight)
• all data points are plotted on the grid
• all data points are connected with a line
• line does not go to origin (lower left corner of grid)

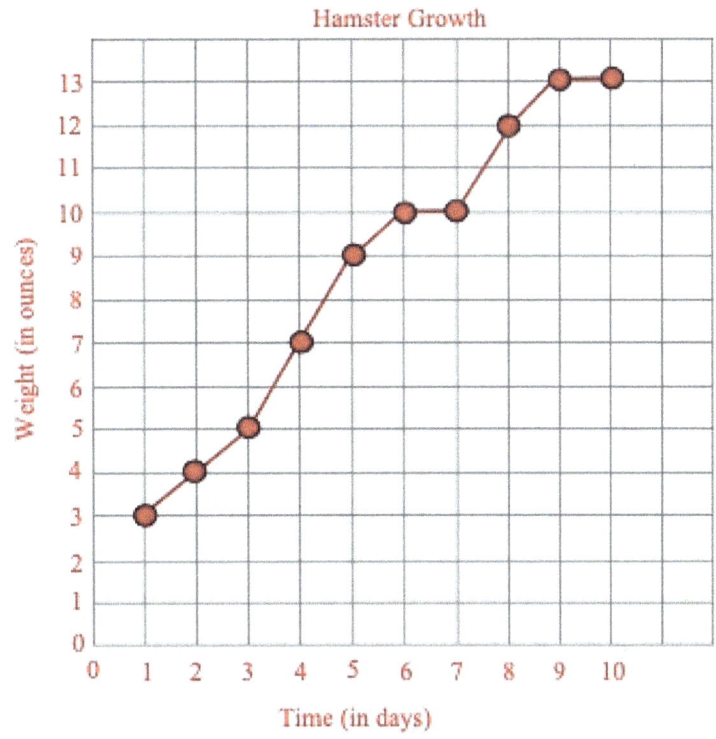

MODULE 5 TEST

THE SCIENCE PROCESS SKILL OF
SPACE-TIME RELATIONSHIPS

Module 5 TEST: SPACE-TIME RELATIONSHIPS

NAME: _____

> **DIRECTIONS:** Read each question carefully. Read each answer carefully. Select the best answer and write its CAPITAL letter in the blank to the left of the question. Each question is worth 1 point unless otherwise noted. Total test value is 17 points.

_____ 1. The amount of positive change in the speed of an object is called . . .
    A. deceleration
    B. acceleration
    C. receleration
    D. disceleration

_____ 2. On the earth, which of the following things can increase the speed of a falling object?
    A. air
    B. water
    C. light
    D. gravity

_____ 3. A negative change in the speed of an object is called . . .
    A. deceleration
    B. acceleration
    C. receleration
    D. disceleration

> Use the following formula for the next question.
>
> $$\text{acceleration} = \frac{(\text{final speed}) - (\text{beginning speed})}{(\text{total time for speed change})}$$

_____ 4. What is the acceleration of a rolling ball that begins with a speed of 10 cm/sec and ends up at a speed of 50 cm/sec and does it in 5 seconds?
    A. 200 cm/s/s
    B. 1 cm/s/s
    C. 8 cm/s/s
    D. 10 cm/s/s

_____ 5. Imagine you have two powders. One has large particles in it and the other has small particles in it. Which of the two powders will be able to go through a funnel fastest?
    A. The one with large particles in it.
    B. The one with small particles in it.
    C. Both will go through the funnel at the same rate.
    D. It depends on the size of the funnel.

GO ON TO NEXT PAGE!

_____ 6. Imagine you want to quickly make a drink using a powdered drink mix. What temperature of water would you use to dissolve the drink mix?
   A. Hot
   B. Room Temperature
   C. Cold
   D. It does not matter

_____ 7. Tiny holes in the bottom of tree leaves that allow moisture and air to go into and out of the leaves are called . . .
   A. transpirers
   B. guard cells
   C. stomates
   D. stomata

_____ 8. When a tree leaf releases water from its leaves, that process is called . . .
   A. transpiration
   B. evaporation
   C. disaporation
   D. dessication

_____ 9. Suppose you have two shirts that got completely wet. Both shirts are made of the same material, but one is twice the size of the other. You hang both shirts on a clothesline to dry them. Which of the two shirts is likely to dry the fastest?
   A. The larger shirt.
   B. The smaller shirt.
   C. Both should dry at the same rate.
   D. Neither will dry if hung together.

_____ 10. Imagine you have two cloth towels and both are completely wet. The towels are the exact same size. One towel is very smooth but the other towel has bumps and ridges on it. You hang both towels on the clothesline to dry. Which of the two towels is likely to dry the fastest?
   A. The smooth one.
   B. The one with bumps and ridges.
   C. Both should dry at the same rate.
   D. Neither will dry if hung together.

_____ 11. The physical space in which an object or person exists is called a . . .
   A. movement space.
   B. objective reference space.
   C. frame of reference.
   D. physical reference.

GO ON TO NEXT PAGE!

_____ 12. The movement or progress of events from the past to the future is called . . .
       A. time
       B. frame of reference
       C. progress reality
       D. gradient

_____ 13. The change in something like temperature or mass or volume over a specific period of time is called a . . .
       A. frame of reference
       B. gradient
       C. parameter
       D. specific reference

_____ 14. The speed at which something changes over time is called a . . .
       A. parameter
       B. specific reference
       C. frame of reference
       D. rate of change

GO ON TO NEXT PAGE!

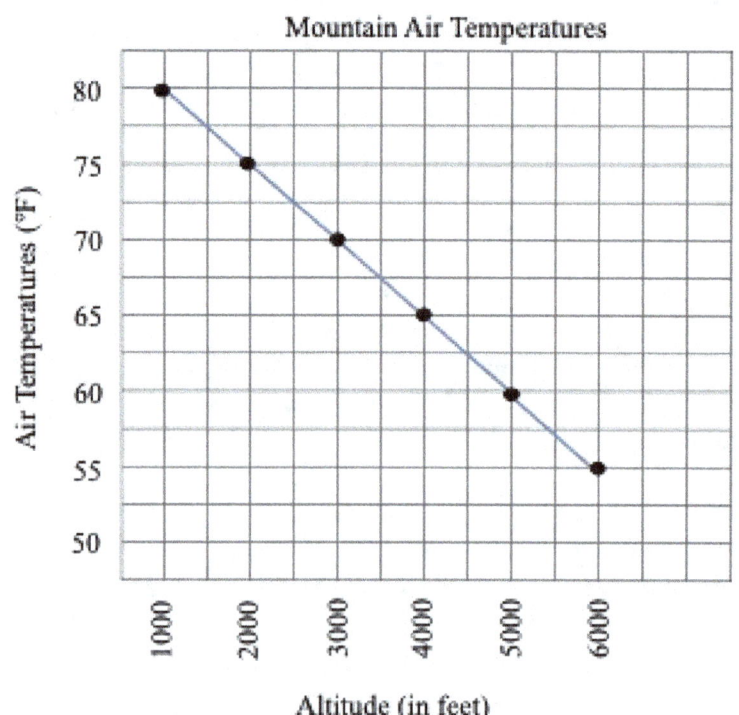

Use the following graph to answer the next 3 questions.

Mountain Air Temperatures

Altitude (in feet)

_____ 15. What happens to the air temperature as altitude increases?
  A. It increases as altitude increases.
  B. It increases as altitude decreases.
  C. It decreases as altitude increases.
  D. It decreases as altitude decreases.

_____ 16. The change in air temperature with the change in altitude is called a . . .
  A. gradient
  B. frame of reference
  C. parameter
  D. physical reference

_____ 17. What can you tell about the air temperature's rate of change as altitude changes?
  A. It increases as altitude increases.
  B. It decreases as altitude increases.
  C. It is the same as altitude decreases.
  D. It is the same as altitude increases.

Module 5 TEST: SPACE-TIME RELATIONSHIPS

NAME: **KEY**

> **DIRECTIONS:** Read each question carefully. Read each answer carefully. Select the best answer and write its CAPITAL letter in the blank to the left of the question. Each question is worth 1 point unless otherwise noted. Total test value is 17 points.

B   1. The amount of positive change in the speed of an object is called . . .
      A. deceleration
      B. acceleration
      C. receleration
      D. disceleration

D   2. On the earth, which of the following things can increase the speed of a falling object?
      A. air
      B. water
      C. light
      D. gravity

A   3. A negative change in the speed of an object is called . . .
      A. deceleration
      B. acceleration
      C. receleration
      D. disceleration

> Use the following formula for the next question.
>
> $$\text{acceleration} = \frac{(\text{final speed}) - (\text{beginning speed})}{(\text{total time for speed change})}$$

C   4. What is the acceleration of a rolling ball that begins with a speed of 10 cm/sec and ends up at a speed of 50 cm/sec and does it in 5 seconds?
      A. 200 cm/s/s
      B. 1 cm/s/s
      C. 8 cm/s/s
      D. 10 cm/s/s

B   5. Imagine you have two powders. One has large particles in it and the other has small particles in it. Which of the two powders will be able to go through a funnel fastest?
      A. The one with large particles in it.
      B. The one with small particles in it.
      C. Both will go through the funnel at the same rate.
      D. It depends on the size of the funnel.

GO ON TO NEXT PAGE!

A 6. Imagine you want to quickly make a drink using a powdered drink mix. What temperature of water would you use to dissolve the drink mix?
  A. Hot
  B. Room Temperature
  C. Cold
  D. It does not matter

D 7. Tiny holes in the bottom of tree leaves that allow moisture and air to go into and out of the leaves are called . . .
  A. transpirers
  B. guard cells
  C. stomates
  D. stomata

A 8. When a tree leaf releases water from its leaves, that process is called . . .
  A. transpiration
  B. evaporation
  C. disaporation
  D. dessication

B 9. Suppose you have two shirts that got completely wet. Both shirts are made of the same material, but one is twice the size of the other. You hang both shirts on a clothesline to dry them. Which of the two shirts is likely to dry the fastest?
  A. The larger shirt.
  B. The smaller shirt.
  C. Both should dry at the same rate.
  D. Neither will dry if hung together.

A 10. Imagine you have two cloth towels and both are completely wet. The towels are the exact same size. One towel is very smooth but the other towel has bumps and ridges on it. You hang both towels on the clothesline to dry. Which of the two towels is likely to dry the fastest?
  A. The smooth one.
  B. The one with bumps and ridges.
  C. Both should dry at the same rate.
  D. Neither will dry if hung together.

C 11. The physical space in which an object or person exists is called a . . .
  A. movement space.
  B. objective reference space.
  C. frame of reference.
  D. physical reference.

GO ON TO NEXT PAGE!

___A___ 12. The movement or progress of events from the past to the future is called . . .
        A. time
        B. frame of reference
        C. progress reality
        D. gradient

___B___ 13. The change in something like temperature or mass or volume over a specific period of time is called a . . .
        A. frame of reference
        B. gradient
        C. parameter
        D. specific reference

___D___ 14. The speed at which something changes over time is called a . . .
        A. parameter
        B. specific reference
        C. frame of reference
        D. rate of change

GO ON TO NEXT PAGE!

Use the following graph to answer the next 3 questions.

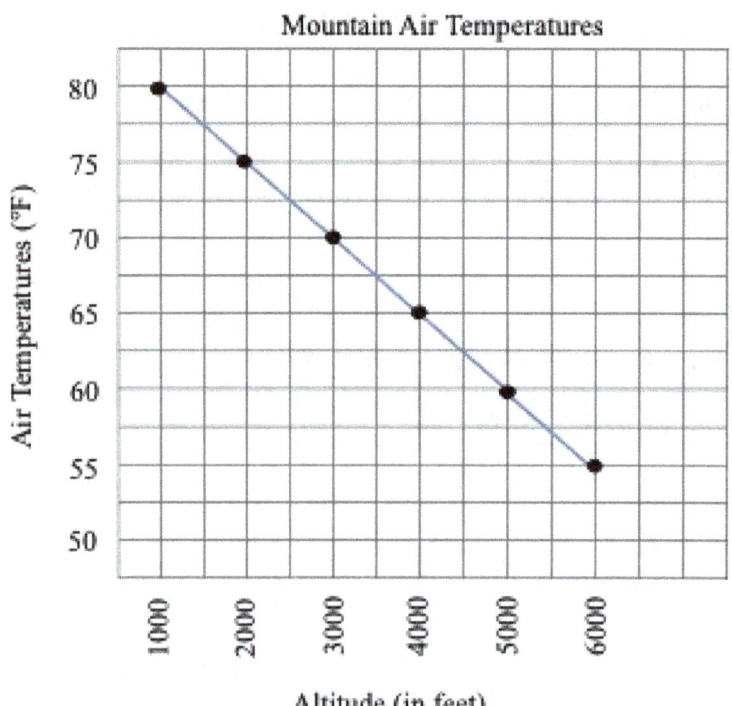

C 15. What happens to the air temperature as altitude increases?
    A. It increases as altitude increases.
    B. It increases as altitude decreases.
    C. It decreases as altitude increases.
    D. It decreases as altitude decreases.

A 16. The change in air temperature with the change in altitude is called a . . .
    A. gradient
    B. frame of reference
    C. parameter
    D. physical reference

D 17. What can you tell about the air temperature's rate of change as altitude changes?
    A. It increases as altitude increases.
    B. It decreases as altitude increases.
    C. It is the same as altitude decreases.
    D. It is the same as altitude increases.

MODULE 6 TEST

THE SCIENCE PROCESS SKILL OF COMMUNICATING

Module 6 TEST: COMMUNICATING

NAME: _____

> **DIRECTIONS:** Read each question carefully. Read each answer carefully. Select the best answer and write its CAPITAL letter in the blank to the left of the question. Each is worth 1 point unless otherwise noted. Total test value is 15 points.

_____ 1. A scientist wanted to determine how many invasive mustard plants were in a field. How could the scientist find out without counting every plant in the field?
  A. Count the mustard plants in several square meters of the field.
  B. Count the mustard plants in a square meter and multiply by the number of square meters in the field.
  C. Have other people come to help count the mustard plants in parts of the field.
  D. Count the mustard plants in one square meter and divide that number by the number of square meters in the field.

_____ 2. If you have 20 sour candies in a bag of candy, and 5 of them are red, how many red sour candies do you think would be in a bigger bag of candy that has 100 candies in it?
  A. 25
  B. 20
  C. 5
  D. 4

_____ 3. What does it mean when we say an operational definition is "context specific"?
  A. It tells what a context is.
  B. It is very specific and precise.
  C. It only applies in a specific situation or at a specific time.
  D. It tells what the operation is that will be done.

4. A student is doing an experiment to find out if one brand of paper towel will dry in the air faster than other brands. The student needs to define what is meant by the paper towel being dry. Write an operational definition for that student's paper towel drying. Write it in the space below. This is worth 3 points.

GO ON TO NEXT PAGE!

5. Look at the diagram below. Write a description of it that you could send to a friend who cannot see it. Write your description so your friend could redraw it. Do not use any sketches in your description. Write your description in the space below the diagram. This is worth 9 points.

Module 6 TEST: COMMUNICATING

NAME: _____**KEY**_____

> **DIRECTIONS:** Read each question carefully. Read each answer carefully. Select the best answer and write its CAPITAL letter in the blank to the left of the question. Each is worth 1 point unless otherwise noted. Total test value is 15 points.

B 1. A scientist wanted to determine how many invasive mustard plants were in a field. How could the scientist find out without counting every plant in the field?
  A. Count the mustard plants in several square meters of the field.
  B. Count the mustard plants in a square meter and multiply by the number of square meters in the field.
  C. Have other people come to help count the mustard plants in parts of the field.
  D. Count the mustard plants in one square meter and divide that number by the number of square meters in the field.

A 2. If you have 20 sour candies in a bag of candy, and 5 of them are red, how many red sour candies do you think would be in a bigger bag of candy that has 100 candies in it?
  A. 25
  B. 20
  C. 5
  D. 4

C 3. What does it mean when we say an operational definition is "context specific"?
  A. It tells what a context is.
  B. It is very specific and precise.
  C. It only applies in a specific situation or at a specific time.
  D. It tells what the operation is that will be done.

4. A student is doing an experiment to find out if one brand of paper towel will dry in the air faster than other brands. The student needs to define what is meant by the paper towel being dry. Write an operational definition for that student's paper towel drying. Write it in the space below. This is worth 3 points.

Score 1 point for each of the following operational definition elements:
• It says what is being defined (e.g. when the paper towel is dry)
• It says how to tell if there is any of it (e.g. the paper towel contains no moisture)
• It tells how to tell how much of it there is (e.g. the paper towel has no wetness, such as it does not feel damp)

GO ON TO NEXT PAGE!

5. Look at the diagram below. Write a description of it that you could send to a friend who cannot see it. Write your description so your friend could redraw it. Do not use any sketches in your description. Write your description in the space below the diagram. This is worth 9 points.

Score one point for each of the following diagram elements:
• one blue triangle with its point up
• one half circle that is white looking like a smile, inside the blue triangle
• two long red rectangles like legs with their outside edges at the outside corners of the blue triangle
• each red rectangle is three times longer than the height of the triangle
• two yellow circles with smaller purple circles inside them
• there is one yellow circle on each side of the blue triangle
• each yellow circle touches its side of the triangle at the middle of the triangle's side
• each yellow circle's diameter is equal to the length of the blue triangle's bottom
• there is no sketch or partial drawing provided in the written description

MODULE 7 TEST

THE SCIENCE PROCESS SKILL OF FORMULATING AND USING MODELS

Module 7 TEST: FORMULATING AND USING MODELS

NAME: _____

DIRECTIONS: Read each question carefully. Read each answer carefully. Select the best answer and write its CAPITAL letter in the blank to the left of the question. Each is worth 1 point unless otherwise noted. Total test value is 20 points.

_____ 1. What is one difference between an insect and a spider?
    A. Insects have 3 body sections and spiders have 2.
    B. Insects have 2 body sections and spiders have 3.
    C. Insects have compound eyes and spiders do not.
    D. Spiders have compound eyes and insects do not.

_____ 2. What is another difference between an insect and a spider?
    A. Insects are smaller than spiders.
    B. Insects have 8 legs and spiders have 6.
    C. Insects have 6 legs and spiders have 8.
    D. Insects are larger than spiders.

_____ 3. A line on a map that connects points that are all the same elevation is called a(an) . . .
    A. connecting line
    B. index line
    C. map line
    D. contour line

_____ 4. On some maps, there are lines of elevation that are darker than the other lines and have elevation numbers on them. These are called . . .
    A. contour indexes
    B. index contours
    C. index lines
    D. key lines

_____ 5. A model does not always have all the details of the thing it represents. This is a
    A. good thing for models
    B. limitation of models
    C. way models are more portable
    D. reason models are not very usable

GO ON TO NEXT PAGE!

_____   6. One advantage of models is they are more . . .
A. portable than the real thing
B. stable than the real thing
C. colorful than the real thing
D. realistic than the real thing

_____   7. Things on a map are usually drawn so they are a certain size compared to the real thing.
An example is one inch on a map equals one mile on the earth. This sizing is called . . .
A. reducing
B. marking down
C. scale
D. gradation

_____   8. The kind of molecular bond that is made when one atom transfers one or more of its
electrons to another atom is called . . .
A. transfer
B. covalent
C. ionic
D. paired

_____   9. The kind of molecular bond that is made when one atom shares electrons with another
atom is called . . .
A. transfer
B. covalent
C. ionic
D. paired

_____   10. The kind of molecular bond that is the strongest is . . .
A. transfer
B. covalent
C. ionic
D. paired

_____   11. The kind of molecular bond that is the weakest is . . .
A. transfer
B. covalent
C. ionic
D. paired

GO ON TO NEXT PAGE!

12. Below is a topographic map. Your task is to make a profile of the topographic map along a line between point A and point B. Fold the bottom of the page so its edge is along a line that would be between point A and point B on the map. Then make the profile on the back of the page that is now folded over. This is worth 7 points.

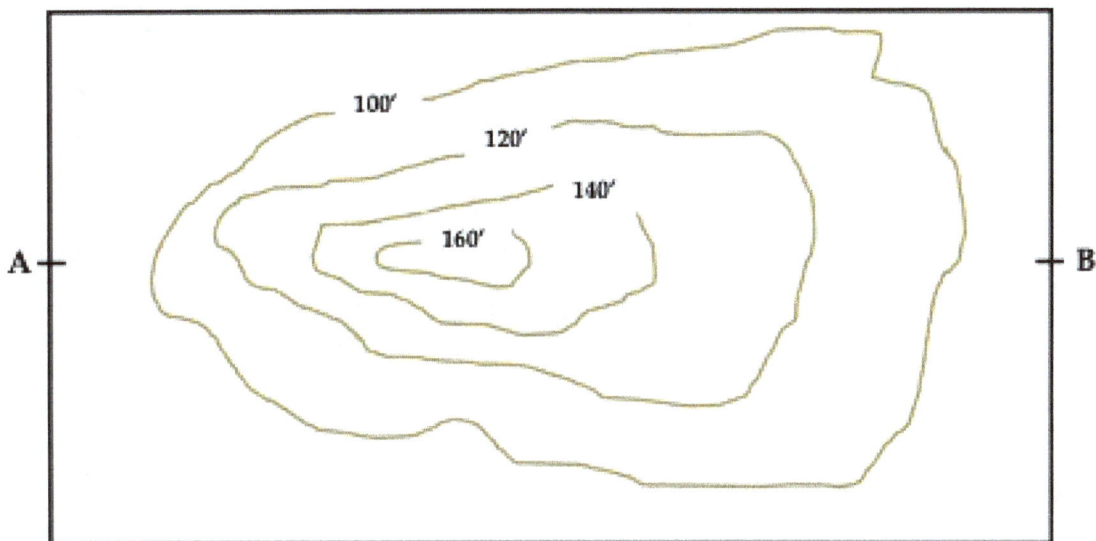

_____ 13. The steepest slope of the landform is on which side of the map?
     A. The middle
     B. The top
     C. The right
     D. The left

_____ 14. The landform shown on the map is a . . .
     A. valley
     B. flat but gently sloping plain
     C. hill
     D. crater

Module 7 TEST: FORMULATING AND USING MODELS

NAME: _____KEY_____

> **DIRECTIONS:** Read each question carefully. Read each answer carefully. Select the best answer and write its CAPITAL letter in the blank to the left of the question. Each is worth 1 point unless otherwise noted. Total test value is 20 points.

_____A_____ 1. What is one difference between an insect and a spider?
   A. Insects have 3 body sections and spiders have 2.
   B. Insects have 2 body sections and spiders have 3.
   C. Insects have compound eyes and spiders do not.
   D. Spiders have compound eyes and insects do not.

_____C_____ 2. What is another difference between an insect and a spider?
   A. Insects are smaller than spiders.
   B. Insects have 8 legs and spiders have 6.
   C. Insects have 6 legs and spiders have 8.
   D. Insects are larger than spiders.

_____D_____ 3. A line on a map that connects points that are all the same elevation is called a(an) . . .
   A. connecting line
   B. index line
   C. map line
   D. contour line

_____B_____ 4. On some maps, there are lines of elevation that are darker than the other lines and have elevation numbers on them. These are called . . .
   A. contour indexes
   B. index contours
   C. index lines
   D. key lines

_____B_____ 5. A model does not always have all the details of the thing it represents. This is a
   A. good thing for models
   B. limitation of models
   C. way models are more portable
   D. reason models are not very usable

GO ON TO NEXT PAGE!

A   6. One advantage of models is they are more . . .
       A. portable than the real thing
       B. stable than the real thing
       C. colorful than the real thing
       D. realistic than the real thing

C   7. Things on a map are usually drawn so they are a certain size compared to the real thing.
       An example is one inch on a map equals one mile on the earth. This sizing is called . . .
       A. reducing
       B. marking down
       C. scale
       D. gradation

C   8. The kind of molecular bond that is made when one atom transfers one or more of its
       electrons to another atom is called . . .
       A. transfer
       B. covalent
       C. ionic
       D. paired

B   9. The kind of molecular bond that is made when one atom shares electrons with another
       atom is called . . .
       A. transfer
       B. covalent
       C. ionic
       D. paired

B   10. The kind of molecular bond that is the strongest is . . .
       A. transfer
       B. covalent
       C. ionic
       D. paired

C   11. The kind of molecular bond that is the weakest is . . .
       A. transfer
       B. covalent
       C. ionic
       D. paired

GO ON TO NEXT PAGE!

12. Below is a topographic map. Your task is to make a profile of the topographic map along a line between point A and point B. Fold the bottom of the page so its edge is along a line that would be between point A and point B on the map. Then make the profile on the back of the page that is now folded over. This is worth 7 points.

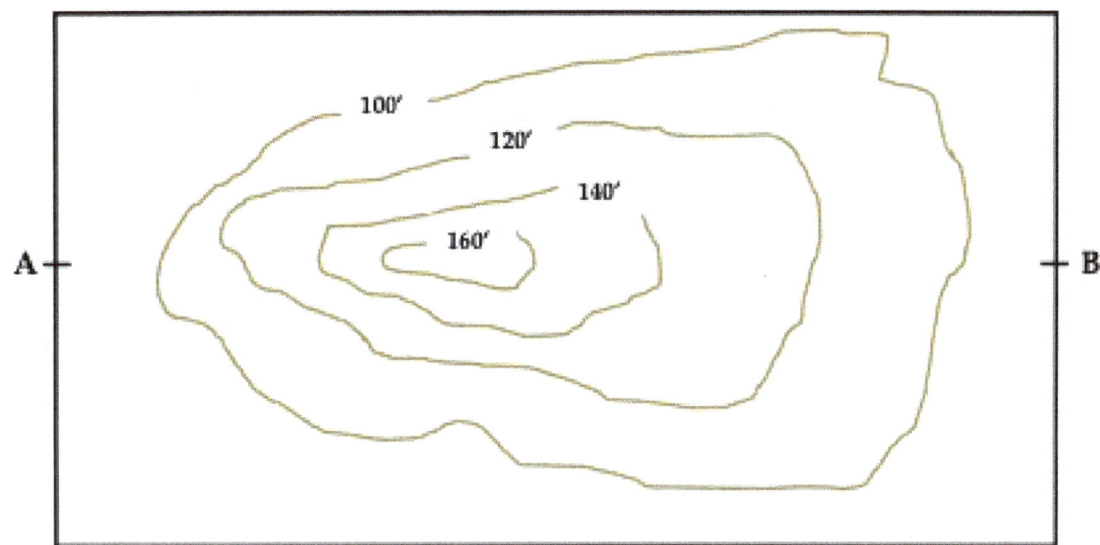

_____ **D** 13. The steepest slope of the landform is on which side of the map?
     A. The middle
     B. The top
     C. The right
     D. The left

_____ **C** 14. The landform shown on the map is a . . .
     A. valley
     B. flat but gently sloping plain
     C. hill
     D. crater

Score one point for each of the following profile elements:
• Edge of paper along A-B line
• Marks matching contour lines at paper edge
• Elevations marked along edge of paper
• Horizontal lines drawn across profile area
• Elevations put in vertical scale
• Elevation points plotted
• Lines connecting plotted points

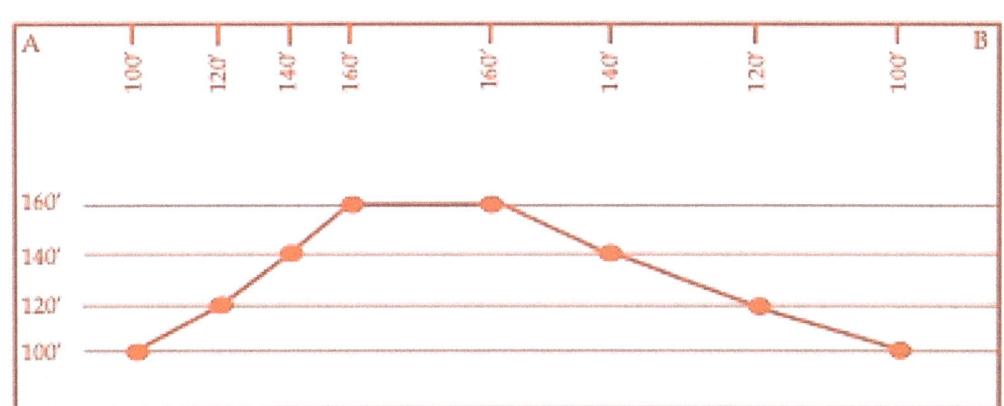

MODULE 8 TEST

THE SCIENCE PROCESS SKILL OF USING NUMBERS

Module 8 TEST: USING NUMBERS

NAME: _____

> **DIRECTIONS:** Read each question carefully. Read each answer carefully. Select the best answer and write its CAPITAL letter in the blank to the left of the question. Each is worth 1 point unless otherwise noted. Total test value is 16 points.

_____ 1. On a compass rose, the direction south (S) is how many degrees?
A. 90°
B. 180°
C. 270°
D. 360°

_____ 2. North, South, East, and West are called . . .
A. compass directives
B. rose directions
C. cardinal directions
D. cardinal points

_____ 3. A key thing to know and be able to find when determining your latitude is . . .
A. where Polaris is
B. how to use a compass rose
C. what cardinal direction you are facing toward
D. also knowing your longitude

_____ 4. When you use an astrolabe to find the altitude angle of the North Star, that angle is the same as . . .
A. the angle on a compass rose
B. the cardinal direction you are facing
C. your longtitude
D. your latitude

_____ 5. An ancient device that can still be used to help you determine your longitude is a(an) . . .
A. compass rose
B. gnomon
C. astrolabe
D. strobe

_____ 6. When determining your longitude, you need to know . . .
A. your time zone
B. cardinal directions
C. your latitude
D. which side of a gnomon is up

GO ON TO NEXT PAGE!

_____ 7. Speed is a derived measurement that includes . . .
      A. weight and time
      B. direction and time
      C. distance and direction
      D. distance and time

_____ 8. The mathematical formula for calculating speed is . . .
      A. $KE = \frac{1}{2}\,mv^2$
      B. $s = t/d$
      C. $s = d/t$
      D. $v = KE/t$

_____ 9. A cart travels a distance of 100 meters in just 25 seconds. What is its speed?
      A. 4 m/s
      B. ¼ m/s
      C. 2500 m/s
      D. 125 m/s

_____ 10. When something moves through a distance in a certain amount of time, and also is moving in a specific direction, it is said to have . . .
      A. speed
      B. velocity
      C. cardinality
      D. compass rosing

_____ 11. The energy of motion is called . . .
      A. motive energy
      B. speed energy
      C. velocity energy
      D. kinetic energy

_____ 12. A distant object viewed from two different locations can appear to shift in its horizontal position when it actually does not move. That apparent movement is called . . .
      A. velocity
      B. kinetic shifting
      C. parallax
      D. repositioning

GO ON TO NEXT PAGE!

_____ 13. As an object that is viewed by someone gets farther and farther away, what will happen to the apparent horizontal shift of its position?
A. it gets smaller
B. it gets larger
C. it does not change
D. it goes vertical

_____ 14. What is a measurement value that is calculated by either multiplying one variable's value times another variable's value, or by dividing one variable's value by another variable's value?
A. variable value
B. multiplied value
C. divided value
D. derived value

_____ 15. The number of degrees of angle between the equator and the north pole (or the south pole) is called . . .
A. polar angle
B. meridian angle
C. latitude
D. longitude

_____ 16. The number of degrees of angle to the west or to the east from the Prime Meridian is called …
A. polar angle
B. meridian angle
C. latitude
D. longitude

Module 8 TEST: USING NUMBERS

NAME: _____ KEY _____

> **DIRECTIONS:** Read each question carefully. Read each answer carefully. Select the best answer and write its CAPITAL letter in the blank to the left of the question. Each is worth 1 point unless otherwise noted. Total test value is 16 points.

B   1. On a compass rose, the direction south (S) is how many degrees?
       A. 90°
       B. 180°
       C. 270°
       D. 360°

C   2. North, South, East, and West are called . . .
       A. compass directives
       B. rose directions
       C. cardinal directions
       D. cardinal points

A   3. A key thing to know and be able to find when determining your latitude is . . .
       A. where Polaris is
       B. how to use a compass rose
       C. what cardinal direction you are facing toward
       D. also knowing your longitude

D   4. When you use an astrolabe to find the altitude angle of the North Star, that angle is the same as . . .
       A. the angle on a compass rose
       B. the cardinal direction you are facing
       C. your longtitude
       D. your latitude

B   5. An ancient device that can still be used to help you determine your longitude is a(an) . . .
       A. compass rose
       B. gnomon
       C. astrolabe
       D. strobe

A   6. When determining your longitude, you need to know . . .
       A. your time zone
       B. cardinal directions
       C. your latitude
       D. which side of a gnomon is up

GO ON TO NEXT PAGE!

D   7. Speed is a derived measurement that includes . . .
- A. weight and time
- B. direction and time
- C. distance and direction
- D. distance and time

C   8. The mathematical formula for calculating speed is . . .
- A. $KE = \frac{1}{2} mv^2$
- B. $s = t/d$
- C. $s = d/t$
- D. $v = KE/t$

A   9. A cart travels a distance of 100 meters in just 25 seconds. What is its speed?
- A. 4 m/s
- B. ¼ m/s
- C. 2500 m/s
- D. 125 m/s

B   10. When something moves through a distance in a certain amount of time, and also is moving in a specific direction, it is said to have . . .
- A. speed
- B. velocity
- C. cardinality
- D. compass rosing

D   11. The energy of motion is called . . .
- A. motive energy
- B. speed energy
- C. velocity energy
- D. kinetic energy

C   12. A distant object viewed from two different locations can appear to shift in its horizontal position when it actually does not move. That apparent movement is called . . .
- A. velocity
- B. kinetic shifting
- C. parallax
- D. repositioning

GO ON TO NEXT PAGE!

_A_ 13. As an object that is viewed by someone gets farther and farther away, what will happen to the apparent horizontal shift of its position?
   A. it gets smaller
   B. it gets larger
   C. it does not change
   D. it goes vertical

_D_ 14. What is a measurement value that is calculated by either multiplying one variable's value times another variable's value, or by dividing one variable's value by another variable's value?
   A. variable value
   B. multiplied value
   C. divided value
   D. derived value

_C_ 15. The number of degrees of angle between the equator and the north pole (or the south pole) is called . . .
   A. polar angle
   B. meridian angle
   C. latitude
   D. longitude

_D_ 16. The number of degrees of angle to the west or to the east from the Prime Meridian is called …
   A. polar angle
   B. meridian angle
   C. latitude
   D. longitude

MODULE 9 TEST

THE SCIENCE PROCESS SKILL OF CLASSIFYING

Module 9 TEST: CLASSIFYING

NAME: _____

> **DIRECTIONS:** Read each question carefully. Read each answer carefully. Select the best answer and write its CAPITAL letter in the blank to the left of the question. Each is worth 1 point unless otherwise noted. Total test value is 9 points.

_____ 1. The grouping objects, events or information based on their properties is . . .
  A. arranging
  B. ordering
  C. classifying
  D. distributing

_____ 2. A system that helps people more easily understand similarities, differences, and relationships between objects and events is . . .
  A. arranging
  B. ordering
  C. classifying
  D. distributing

_____ 3. Attributes and attribute values of objects or events are important when doing . . .
  A. arranging
  B. ordering
  C. classifying
  D. distributing

_____ 4. A system to place items into one group or another group, either having a certain attribute or not having it, is called . . .
  A. dichotomous pairing
  B. dichotomous classification
  C. binary grouping
  D. paired attributing

_____ 5. When you relate a small group of things back to the larger group to which it belongs, that process is called . . .
  A. regrouping
  B. class dichotomizing
  C. attribute pairing
  D. class inclusion

GO ON TO NEXT PAGE!

_____ 6. According to the Linnean system, the largest grouping of things is called a . . .
A. kingdom
B. phylum
C. class
D. species

_____ 7. According to the Linnean system, the smallest grouping of things is called a . . .
A. kingdom
B. phylum
C. class
D. species

_____ 8. To keep things simple in the Linnean system, the two group names that are usually used for something are . . .
A. kingdom and phylum
B. family and species
C. class and order
D. genus and species

_____ 9. If a classification system is to be effectively used, people need to understand . . .
A. the purpose for the system.
B. how to work the system forward and backward.
C. what things need to be excluded from the system.
D. how to use Latin for the names of things.

Module 9 TEST: CLASSIFYING

NAME: _____KEY_____

DIRECTIONS: Read each question carefully. Read each answer carefully. Select the best answer and write its CAPITAL letter in the blank to the left of the question. Each is worth 1 point unless otherwise noted. Total test value is 9 points.

_C_ 1. The grouping objects, events or information based on their properties is . . .
  A. arranging
  B. ordering
  C. classifying
  D. distributing

_C_ 2. A system that helps people more easily understand similarities, differences, and relationships between objects and events is . . .
  A. arranging
  B. ordering
  C. classifying
  D. distributing

_C_ 3. Attributes and attribute values of objects or events are important when doing . . .
  A. arranging
  B. ordering
  C. classifying
  D. distributing

_B_ 4. A system to place items into one group or another group, either having a certain attribute or not having it, is called . . .
  A. dichotomous pairing
  B. dichotomous classification
  C. binary grouping
  D. paired attributing

_D_ 5. When you relate a small group of things back to the larger group to which it belongs, that process is called . . .
  A. regrouping
  B. class dichotomizing
  C. attribute pairing
  D. class inclusion

GO ON TO NEXT PAGE!

_____A_____ 6. According to the Linnean system, the largest grouping of things is called a . . .
A. kingdom
B. phylum
C. class
D. species

_____D_____ 7. According to the Linnean system, the smallest grouping of things is called a . . .
A. kingdom
B. phylum
C. class
D. species

_____D_____ 8. To keep things simple in the Linnean system, the two group names that are usually used for something are . . .
A. kingdom and phylum
B. family and species
C. class and order
D. genus and species

_____A_____ 9. If a classification system is to be effectively used, people need to understand . . .
A. the purpose for the system.
B. how to work the system forward and backward.
C. what things need to be excluded from the system.
D. how to use Latin for the names of things.

MODULE 10 TEST

THE SCIENCE PROCESS SKILL OF IDENTIFYING AND CONTROLLING VARIABLES

Module 10 TEST: IDENTIFYING AND CONTROLLING VARIABLES

NAME: _____

> **DIRECTIONS:** Read each question carefully. Read each answer carefully. Select the best answer and write its CAPITAL letter in the blank to the left of the question. Each is worth 1 point unless otherwise noted. Total test value is 25 points.

_____ 1. A variable is something that . . .
   A. alters something
   B. can change
   C. must be controlled
   D. should be saved

_____ 2. The type of variable that you change or manipulate is called a(an) . . .
   A. controlled variable
   B. extraneous variable
   C. dependent variable
   D. independent variable

_____ 3. The type of variable that responds to or changes because of something else is called a . . .
   A. controlled variable
   B. extraneous variable
   C. dependent variable
   D. independent variable

_____ 4. The type of variable that needs to remain the same (or is kept constant) through an investigation is called a . . .
   A. controlled variable
   B. extraneous variable
   C. dependent variable
   D. independent variable

_____ 5. A variable that is beyond our control but still affects everything else about the same is called a(an) . . .
   A. controlled variable
   B. extraneous variable
   C. dependent variable
   D. independent variable

GO ON TO NEXT PAGE!

_____ 6. When we set up an investigation or experiment, the number of variables we want to allow to change is(are) . . .
A. 1
B. 2
C. 3
D. 4 or more

_____ 7. During an investigation or experiment, we usually try to allow only one variable to change at a time. This is called . . .
A. maintaining variables
B. constraining variables
C. manipulating variables
D. controlling variables

_____ 8. In an investigation or experiment, we often decide on one set of conditions that we think is the most normal one. This condition is called the . . .
A. normal control
B. experimental control
C. investigative control
D. normed variable

_____ 9. Why is it a problem to allow more than one variable to be changed in an investigation or experiment?
A. It can get too confusing to keep track of all the variables.
B. It causes the number of extraneous variables to increase.
C. We can't know which variable caused the change in the results.
D. There is a limit to how many controls we can set up.

GO TO THE NEXT PAGE!

Use the following diagram and story to answer the next 16 questions.

A student wanted to find out if different colors of light would affect the growth of a plant. The student decided to set up an experiment to try to find out. The student got three plants and put one under red light only, one under blue light only, and one in regular sunlight.

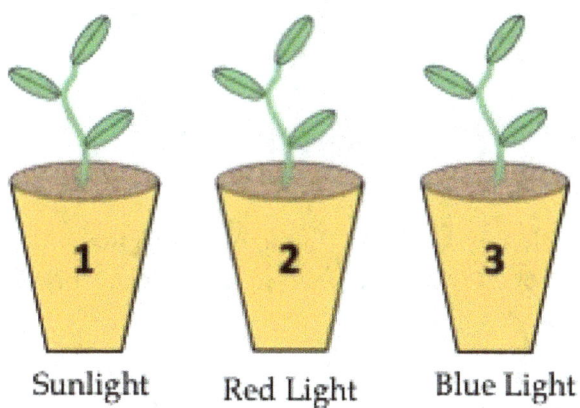

Sunlight   Red Light   Blue Light

_____ 10. In this experiment, what is the independent variable?
    A. The plant in the sunlight
    B. The color of the light
    C. The growth of the plant
    D. The number of plants

_____ 11. In this experiment, what is the dependent variable?
    A. The plant in the sunlight
    B. The color of the light
    C. The growth of the plant
    D. The number of plants

_____ 12. In this experiment, what is the experimental control?
    A. The plant in the sunlight
    B. The color of the light
    C. The growth of the plant
    D. The number of plants

GO ON TO THE NEXT PAGE!

Sunlight   Red Light   Blue Light

DIRECTIONS: The following is a list of variables. In the blanks to the left of each one, write a "**C**" if it should be a **controlled variable** and write an "**N**" (for "**NO**") if it should not be a controlled variable.

_____ 13. amount of plant growth

_____ 14. size of plant pots

_____ 15. amount of gravity pulling on the plants

_____ 16. amount of water used on each plant

_____ 17. kind of soil used for each plant

_____ 18. amount of air around the plants

_____ 19. kind of plant food used

_____ 20. type of plant used

_____ 21. color of light on the plants

_____ 22. the number written on the plant pot

_____ 23. amount of plant food for each plant

_____ 24. temperature where each plant is placed

_____ 25. the number of people measuring the plants

Module 10 TEST: IDENTIFYING AND CONTROLLING VARIABLES

NAME: KEY

DIRECTIONS: Read each question carefully. Read each answer carefully. Select the best answer and write its CAPITAL letter in the blank to the left of the question. Each is worth 1 point unless otherwise noted. Total test value is 25 points.

B   1. A variable is something that . . .
- A. alters something
- B. can change
- C. must be controlled
- D. should be saved

D   2. The type of variable that you change or manipulate is called a(an) . . .
- A. controlled variable
- B. extraneous variable
- C. dependent variable
- D. independent variable

C   3. The type of variable that responds to or changes because of something else is called a . . .
- A. controlled variable
- B. extraneous variable
- C. dependent variable
- D. independent variable

A   4. The type of variable that needs to remain the same (or is kept constant) through an investigation is called a . . .
- A. controlled variable
- B. extraneous variable
- C. dependent variable
- D. independent variable

B   5. A variable that is beyond our control but still affects everything else about the same is called a(an) . . .
- A. controlled variable
- B. extraneous variable
- C. dependent variable
- D. independent variable

GO ON TO NEXT PAGE!

_____A_____ 6. When we set up an investigation or experiment, the number of variables we want to allow to change is(are) . . .
A. 1
B. 2
C. 3
D. 4 or more

_____D_____ 7. During an investigation or experiment, we usually try to allow only one variable to change at a time. This is called . . .
A. maintaining variables
B. constraining variables
C. manipulating variables
D. controlling variables

_____B_____ 8. In an investigation or experiment, we often decide on one set of conditions that we think is the most normal one. This condition is called the . . .
A. normal control
B. experimental control
C. investigative control
D. normed variable

_____C_____ 9. Why is it a problem to allow more than one variable to be changed in an investigation or experiment?
A. It can get too confusing to keep track of all the variables.
B. It causes the number of extraneous variables to increase.
C. We can't know which variable caused the change in the results.
D. There is a limit to how many controls we can set up.

GO TO THE NEXT PAGE!

> Use the following diagram and story to answer the next 16 questions.

A student wanted to find out if different colors of light would affect the growth of a plant. The student decided to set up an experiment to try to find out. The student got three plants and put one under red light only, one under blue light only, and one in regular sunlight.

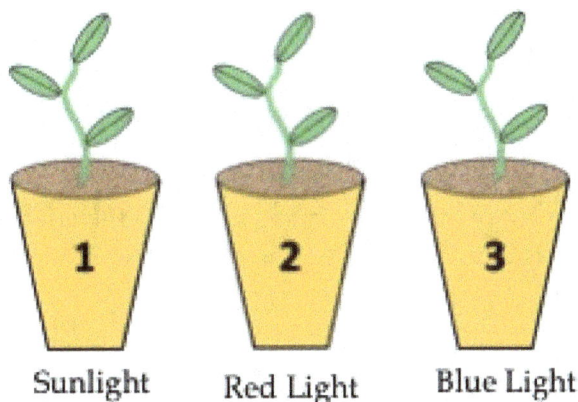

1	2	3
Sunlight	Red Light	Blue Light

B 10. In this experiment, what is the independent variable?
 A. The plant in the sunlight
 B. The color of the light
 C. The growth of the plant
 D. The number of plants

C 11. In this experiment, what is the dependent variable?
 A. The plant in the sunlight
 B. The color of the light
 C. The growth of the plant
 D. The number of plants

A 12. In this experiment, what is the experimental control?
 A. The plant in the sunlight
 B. The color of the light
 C. The growth of the plant
 D. The number of plants

GO ON TO THE NEXT PAGE!

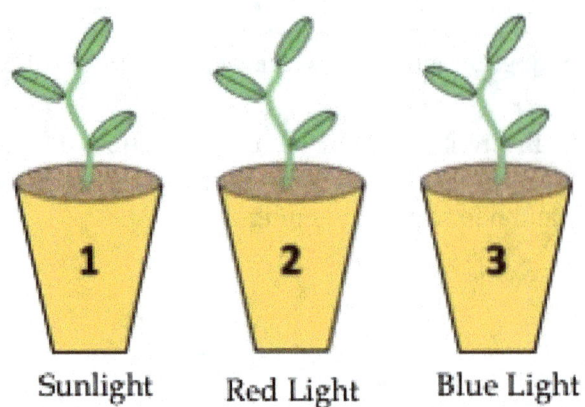

Sunlight   Red Light   Blue Light

DIRECTIONS: The following is a list of variables. In the blanks to the left of each one, write a "**C**" if it should be a **Controlled variable** and write an "**N**" (for "**NO**") if it should not be a controlled variable.

___N___   13.   amount of plant growth

___C___   14.   size of plant pots

___N___   15.   amount of gravity pulling on the plants

___C___   16.   amount of water used on each plant

___C___   17.   kind of soil used for each plant

___N___   18.   amount of air around the plants

___C___   19.   kind of plant food used

___C___   20.   type of plant used

___N___   21.   color of light on the plants

___N___   22.   the number written on the plant pot

___C___   23.   amount of plant food for each plant

___C___   24.   temperature where each plant is placed

___N___   25.   the number of people measuring the plants

MODULE 11 TEST

THE SCIENCE PROCESS SKILL OF QUESTIONING

Module 11 TEST: QUESTIONING

NAME: _____

DIRECTIONS: Read each question carefully. Read each answer carefully. Select the best answer and write its CAPITAL letter in the blank to the left of the question. Each is worth 1 point unless otherwise noted. Total test value is 8 points.

_____ 1. A good scientific question needs to focus on a certain idea. So, a good scientific question needs to be . . .
A. specific
B. measurable
C. achievable
D. relevant
E. time-based

_____ 2. A good scientific question needs to be something that will allow someone to collect observations and data through scientific methods. So, a good scientific question needs to be . . .
A. time-based
B. measurable
C. relevant
D. specific
E. achievable

_____ 3. A good scientific question needs to be something about which someone is able to find an answer. So, a good scientific question needs to be . . .
A. achievable
B. measurable
C. specific
D. time-based
E. relevant

_____ 4. A good scientific question needs to be something that can help someone expand their knowledge about something. So, a good scientific question needs to be . . .
A. specific
B. measurable
C. time-based
D. specific
E. relevant

GO ON TO THE NEXT PAGE!

_____ 5. A good scientific question needs to be something that can be answerable in a specified amount of time. So, a good scientific question needs to be . . .
   A. relevant
   B. specific
   C. measurable
   D. time-based
   E. achievable

_____ 6. A good scientific question can lead to . . .
   A. getting all the answers
   B. puzzling results
   C. formation of a testable hypothesis
   D. freedom of choice

_____ 7. In what way does adding something to water affect the water's surface tension?
   A. It increases it.
   B. It decreases it.
   C. It allows the water to stay the same.
   D. It eliminates it.

_____ 8. If water has a lot of surface tension and a drop of it is put on a table top, the drop will . . .
   A. form a rounded top
   B. become less rounded
   C. flatten out
   D. spread out over the table top

Module 11 TEST: QUESTIONING

NAME: <u>   KEY   </u>

DIRECTIONS: Read each question carefully. Read each answer carefully. Select the best answer and write its CAPITAL letter in the blank to the left of the question. Each is worth 1 point unless otherwise noted. Total test value is 8 points.

<u>A</u>   1. A good scientific question needs to focus on a certain idea. So, a good scientific question needs to be . . .
- A. specific
- B. measurable
- C. achievable
- D. relevant
- E. time-based

<u>B</u>   2. A good scientific question needs to be something that will allow someone to collect observations and data through scientific methods. So, a good scientific question needs to be . . .
- A. time-based
- B. measurable
- C. relevant
- D. specific
- E. achievable

<u>A</u>   3. A good scientific question needs to be something about which someone is able to find an answer. So, a good scientific question needs to be . . .
- A. achievable
- B. measurable
- C. specific
- D. time-based
- E. relevant

<u>E</u>   4. A good scientific question needs to be something that can help someone expand their knowledge about something. So, a good scientific question needs to be . . .
- A. specific
- B. measurable
- C. time-based
- D. specific
- E. relevant

GO ON TO THE NEXT PAGE!

___D___ 5. A good scientific question needs to be something that can be answerable in a specified
amount of time. So, a good scientific question needs to be . . .
A. relevant
B. specific
C. measurable
D. time-based
E. achievable

___C___ 6. A good scientific question can lead to . . .
A. getting all the answers
B. puzzling results
C. formation of a testable hypothesis
D. freedom of choice

___B___ 7. In what way does adding something to water affect the water's surface tension?
A. It increases it.
B. It decreases it.
C. It allows the water to stay the same.
D. It eliminates it.

___A___ 8. If water has a lot of surface tension and a drop of it is put on a table top, the drop will . . .
A. form a rounded top
B. become less rounded
C. flatten out
D. spread out over the table top

MODULE 12 TEST

THE SCIENCE PROCESS SKILL OF HYPOTHESIZING

Module 12 TEST: HYPOTHESIZING

NAME: _____

> DIRECTIONS: Read each question carefully. Read each answer carefully. Select the best answer and write its CAPITAL letter in the blank to the left of the question. Each is worth 1 point unless otherwise noted. Total test value is 14 points.

_____ 1. A statement about what will happen to a single variable is called a . . .
A. singularity
B. prediction
C. scientific question
D. hypothesis

_____ 2. A statement about what will happen when two or more variables interact is called a . . .
A. dualarity
B. prediction
C. scientific question
D. hypothesis

_____ 3. A statement that is something like, "If X occurs, then Y will occur" is a . . .
A. scientific question
B. prediction
C. hypothesis
D. problem

_____ 4. A statement that is something like, "What would happen if . . ." is a . . .
A. problem
B. hypothesis
C. prediction
D. scientific question

_____ 5. What can you say about a hypothesis if it is supported by the results of an investigation or experiment?
A. It is proven
B. It is supported
C. It is predicted
D. It is established

GO ON TO THE NEXT PAGE!

_____ 6. If a hypothesis is not supported by the results of an investigation or experiment, then that hypothesis could be . . .
A. discarded and a new one made.
B. kept and used again.
C. considered to be false.
D. not sophisticated enough for what is being tested.

_____ 7. If a hypothesis is not supported by the results of an investigation or experiment, then that hypothesis could be . . .
A. considered as okay to keep using as it is.
B. ignored as untrue.
C. misleading the investigation.
D. revised and tested again.

_____ 8. What should be the *first step* to do if the results of an investigation or experiment do not support a hypothesis?
A. The hypothesis should be revised and then used in the next investigation.
B. The tests should be done again to see if the investigator made errors.
C. The hypothesis should be kept despite the results of the investigation.
D. The hypothesis should be discarded and a new one made that can be tested.

_____ 9. What should be the *last step* to do if a hypothesis is not supported by the results of an investigation or experiment?
A. The hypothesis should be revised and then used in the next investigation.
B. The tests should be done again to see if the investigator made errors.
C. The hypothesis should be kept despite the results of the investigation.
D. The hypothesis should be discarded and a new one made that can be tested.

_____ 10. What can you say about a hypothesis that *is supported* by the results of an investigation or experiment?
A. it is proven
B. it is disproven
C. it is supported
D. it is not supported

_____ 11. What can you say about a hypothesis that *is not supported* by the results of an investigation or experiment?
A. it is proven
B. it is disproven
C. it is supported
D. it is not supported

GO ON TO THE NEXT PAGE!

> Use the following story to answer the next 3 questions.

A scientist wants to test the whitening power of toothpastes to see which one is the best. The scientist gets five different brands of toothpaste:

- Teeth Gleam Toothpaste
- Shiny Ivories Toothpaste
- Smiley White Toothpaste
- Power Brush Toothpaste
- Zappo Toothpaste

The scientist begins to think about how to conduct the tests. The scientist thinks Teeth Gleam Toothpaste will get rid of the most stains on teeth. The scientist then does the tests on teeth that have coffee stains on them.

_____ 12. A good hypothesis would be . . .
A. Teeth Gleam Toothpaste will be better than the other brands of toothpaste.
B. If a toothpaste is a gel, then it will be the best of the toothpaste brands.
C. If a toothpaste removes the most stains, then it has the most whitening power.
D. Smiley White Toothpaste will whiten teeth better than any brand of toothpaste.

_____ 13. The scientist tests the five toothpastes and finds that Shiny Ivories Toothpaste removed more stains than the other toothpastes. That result did not match the hypothesis the scientist made at the beginning. What can the scientist conclude about the toothpaste test?
A. A mistake must have been made in the testing procedures.
B. The hypothesis was not supported.
C. The hypothesis was completely wrong.
D. The hypothesis is correct but the toothpaste was wrong.

_____ 14. The scientist finds out that Shiny Ivories Toothpaste cleans better than any of the other brands of toothpastes. What can the scientist conclude about the hypothesis of the study?
A. Shiny Ivories Toothpaste is the best toothpaste available to buy.
B. Shiny Ivories Toothpaste happened to be better for coffee stains.
C. Shiny Ivories Toothpaste is the best toothpaste for any stains.
D. The hypothesis needs to be revised for further testing.

Module 12 TEST: HYPOTHESIZING

NAME: _____KEY_____

DIRECTIONS: Read each question carefully. Read each answer carefully. Select the best answer and write its CAPITAL letter in the blank to the left of the question. Each is worth 1 point unless otherwise noted. Total test value is 14 points.

B   1.   A statement about what will happen to a single variable is called a . . .
   A. singularity
   B. prediction
   C. scientific question
   D. hypothesis

D   2.   A statement about what will happen when two or more variables interact is called a . . .
   A. dualarity
   B. prediction
   C. scientific question
   D. hypothesis

C   3.   A statement that is something like, "If X occurs, then Y will occur" is a . . .
   A. scientific question
   B. prediction
   C. hypothesis
   D. problem

B   4.   A statement that is something like, "What would happen if . . ." is a . . .
   A. problem
   B. hypothesis
   C. prediction
   D. scientific question

B   5.   What can you say about a hypothesis if it is supported by the results of an investigation or experiment?
   A. It is proven
   B. It is supported
   C. It is predicted
   D. It is established

GO ON TO THE NEXT PAGE!

A   6.   If a hypothesis is not supported by the results of an investigation or experiment, then that hypothesis could be . . .
A.   discarded and a new one made.
B.   kept and used again.
C.   considered to be false.
D.   not sophisticated enough for what is being tested.

D   7.   If a hypothesis is not supported by the results of an investigation or experiment, then that hypothesis could be . . .
A.   considered as okay to keep using as it is.
B.   ignored as untrue.
C.   misleading the investigation.
D.   revised and tested again.

B   8.   What should be the *first step* to do if the results of an investigation or experiment do not support a hypothesis?
A.   The hypothesis should be revised and then used in the next investigation.
B.   The tests should be done again to see if the investigator made errors.
C.   The hypothesis should be kept despite the results of the investigation.
D.   The hypothesis should be discarded and a new one made that can be tested.

D   9.   What should be the *last step* to do if a hypothesis is not supported by the results of an investigation or experiment?
A.   The hypothesis should be revised and then used in the next investigation.
B.   The tests should be done again to see if the investigator made errors.
C.   The hypothesis should be kept despite the results of the investigation.
D.   The hypothesis should be discarded and a new one made that can be tested.

C   10.   What can you say about a hypothesis that *is supported* by the results of an investigation or experiment?
A.   it is proven
B.   it is disproven
C.   it is supported
D.   it is not supported

B   11.   What can you say about a hypothesis that *is not supported* by the results of an investigation or experiment?
A.   it is proven
B.   it is disproven
C.   it is supported
D.   it is not supported

GO ON TO THE NEXT PAGE!

Use the following story to answer the next 3 questions.

A scientist wants to test the whitening power of toothpastes to see which one is the best. The scientist gets five different brands of toothpaste:

- Teeth Gleam Toothpaste
- Shiny Ivories Toothpaste
- Smiley White Toothpaste
- Power Brush Toothpaste
- Zappo Toothpaste

The scientist begins to think about how to conduct the tests. The scientist thinks Teeth Gleam Toothpaste will get rid of the most stains on teeth. The scientist then does the tests on teeth that have coffee stains on them.

C   12.   A good hypothesis would be . . .
A. Teeth Gleam Toothpaste will be better than the other brands of toothpaste.
B. If a toothpaste is a gel, then it will be the best of the toothpaste brands.
C. If a toothpaste removes the most stains, then it has the most whitening power.
D. Smiley White Toothpaste will whiten teeth better than any brand of toothpaste.

B   13.   The scientist tests the five toothpastes and finds that Shiny Ivories Toothpaste removed more stains than the other toothpastes. That result did not match the hypothesis the scientist made at the beginning. What can the scientist conclude about the toothpaste test?
A. A mistake must have been made in the testing procedures.
B. The hypothesis was not supported.
C. The hypothesis was completely wrong.
D. The hypothesis is correct but the toothpaste was wrong.

D   14.   The scientist finds out that Shiny Ivories Toothpaste cleans better than any of the other brands of toothpastes. What can the scientist conclude about the hypothesis of the study?
A. Shiny Ivories Toothpaste is the best toothpaste available to buy.
B. Shiny Ivories Toothpaste happened to be better for coffee stains.
C. Shiny Ivories Toothpaste is the best toothpaste for any stains.
D. The hypothesis needs to be revised for further testing.

MODULE 13 TEST

THE SCIENCE PROCESS SKILL OF PREDICTING

Module 13 TEST: PREDICTING

NAME: _____

> DIRECTIONS: Read each question carefully. Read each answer carefully. Select the best answer and write its CAPITAL letter in the blank to the left of the question. Each is worth 1 point unless otherwise noted. Total test value is 14 points.

_____ 1. A statement about what will happen to a single variable is called a . . .
   A. singularity
   B. prediction
   C. scientific question
   D. hypothesis

_____ 2. A statement about what will happen when two or more variables interact is called a . . .
   A. dualarity
   B. prediction
   C. scientific question
   D. hypothesis

_____ 3. Using observations to make reasonable guesses about what will happen with something is a . . .
   A. prediction
   B. observational guess
   C. hypothetical guess
   D. reasonable determination

_____ 4. A good kind of question to ask yourself when trying to make a prediction is . . .
   A. "Has this been done before?"
   B. "What materials do I need in order to test it?"
   C. "What do I think will happen if . . . ?"
   D. "Can this be repeated?"

_____ 5. Better quality predictions are made when we have available . . .
   A. multiple observations about something.
   B. good equipment for testing things.
   C. multiple people deciding on what predictions should say.
   D. enough time to think through scientific concepts.

GO ON TO THE NEXT PAGE!

Use the following diagram to answer the next 6 questions. The diagram shows you standing on the playground looking at your shadows. The sun is to your back at this time of the day and is casting your shadow in front of you.

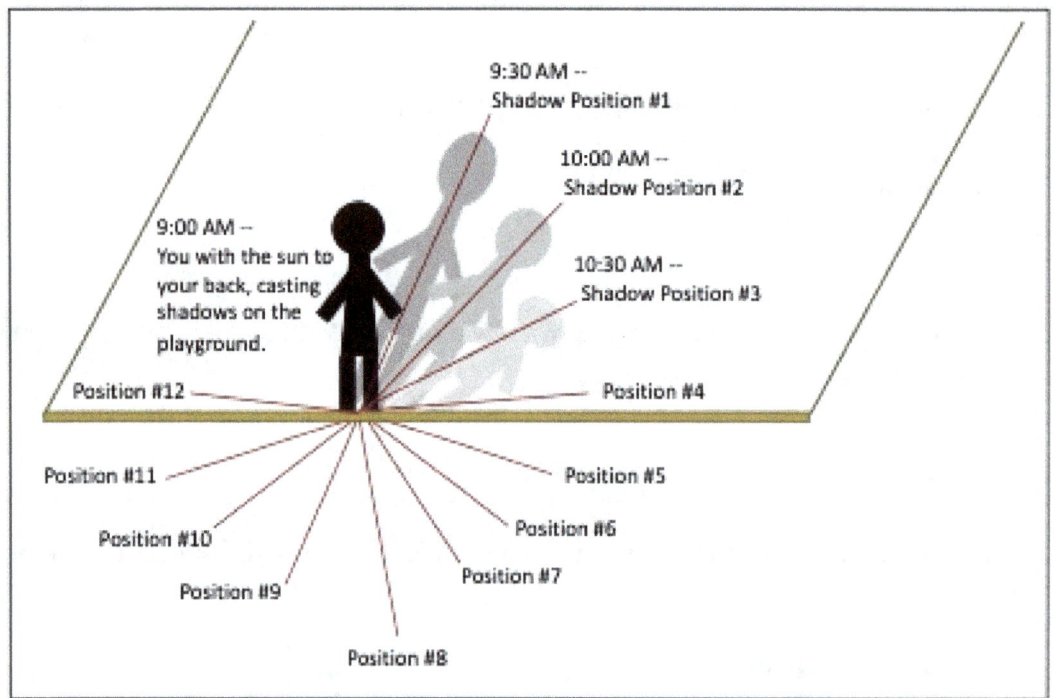

____ 6. What prediction can you make about your shadow at 11:00 AM?
   A. It will be longer and at Position #4
   B. It will be shorter and at Position #4
   C. It will be longer and at Position #12
   D. It will be shorter and at Position #12

____ 7. How tall do you predict your shadow will be at 11:00 AM?
   A. shorter than at Shadow Position #3
   B. longer than at Shadow Position #3
   C. the same length as the one at Shadow Position #3
   D. there will be no shadow at that Shadow Position

____ 8. What Shadow Position # do you predict your shadow will be at noon?
   A. 5
   B. 6
   C. 7
   D. 8

GO ON TO THE NEXT PAGE!

_____ 9. At which shadow position # do you predict your shadow will again be as long as it is at Shadow Position #1?
A. 10
B. 9
C. 8
D. 7

_____ 10. What do you predict will happen to the length of your shadows as the sun goes from the morning into the afternoon?
A. They will get longer at first and then shorter
B. They will shorter at first and then longer
C. They will get shorter each half hour
D. They will get longer each half hour

_____ 11. What do you predict will happen to the positions of your shadows as the sun goes from the morning into the afternoon?
A. They will move into positions #5, then #6, then #7, and then #8
B. They will move into positions #9, then #10, then #11, and then #12
C. They will move into positions #12, then #11, then #10, and then #9
D. They will move into positions #8, then #7, then #6, and then #5

GO ON TO THE NEXT PAGE!

Use the following diagram to answer the next 3 questions. The diagram is a graph showing the time candles will burn in jars of differing sizes or volumes.

CANDLES AND JARS: BURN TIME vs VOLUME

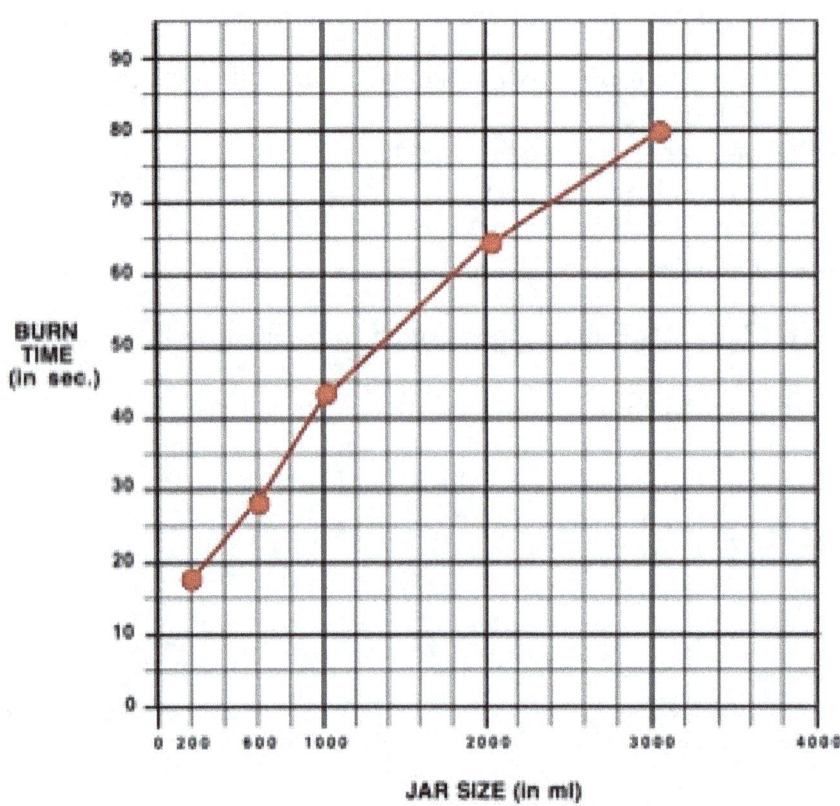

_____ 12. What do you predict the burn time of a candle will be in a jar that has a volume of 800 ml?
A. 35 seconds
B. 38 seconds
C. 40 seconds

_____ 13. What do you predict will be the burn time of a candle in a 4,000 ml jar?
A. it cannot be determined
B. 80 seconds
C. 95 seconds
D. 90 seconds

_____ 14. What do you predict the volume of a jar would be if a candle's burn time in it is 60 seconds?
A. 1800 ml
B. 1500 ml
C. 1000 ml
D. 2000 ml

Module 13 TEST: PREDICTING

NAME: _____KEY_____

DIRECTIONS: Read each question carefully. Read each answer carefully. Select the best answer and write its CAPITAL letter in the blank to the left of the question. Each is worth 1 point unless otherwise noted. Total test value is 14 points.

B   1.   A statement about what will happen to a single variable is called a . . .
A. singularity
B. prediction
C. scientific question
D. hypothesis

D   2.   A statement about what will happen when two or more variables interact is called a . . .
A. dualarity
B. prediction
C. scientific question
D. hypothesis

A   3.   Using observations to make reasonable guesses about what will happen with something is a . . .
A. prediction
B. observational guess
C. hypothetical guess
D. reasonable determination

C   4.   A good kind of question to ask yourself when trying to make a prediction is . . .
A. "Has this been done before?"
B. "What materials do I need in order to test it?"
C. "What do I think will happen if . . . ?"
D. "Can this be repeated?"

A   5.   Better quality predictions are made when we have available . . .
A. multiple observations about something.
B. good equipment for testing things.
C. multiple people deciding on what predictions should say.
D. enough time to think through scientific concepts.

GO ON TO THE NEXT PAGE!

Use the following diagram to answer the next 6 questions. The diagram shows you standing on the playground looking at your shadows. The sun is to your back at this time of the day and is casting your shadow in front of you.

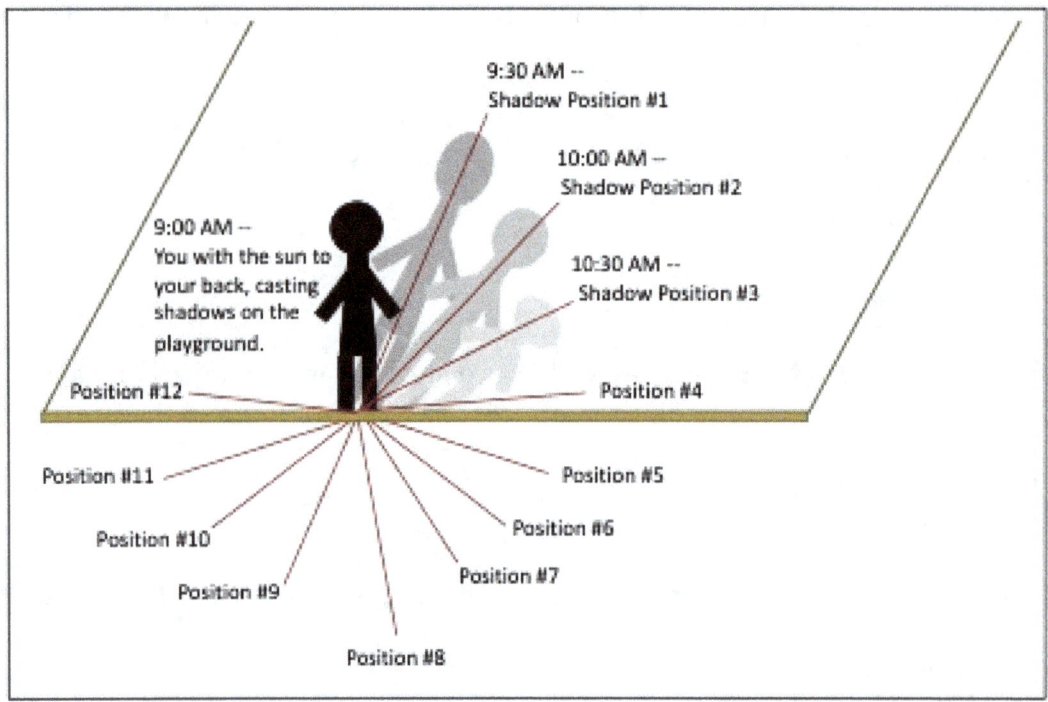

_B_   6. What prediction can you make about your shadow at 11:00 AM?
A. It will be longer and at Position #4
B. It will be shorter and at Position #4
C. It will be longer and at Position #12
D. It will be shorter and at Position #12

_A_   7. How tall do you predict your shadow will be at 11:00 AM?
A. shorter than at Shadow Position #3
B. longer than at Shadow Position #3
C. the same length as the one at Shadow Position #3
D. there will be no shadow at that Shadow Position

_B_   8. What Shadow Position # do you predict your shadow will be at noon?
A. 5
B. 6
C. 7
D. 8

GO ON TO THE NEXT PAGE!

_C_ 9. At which shadow position # do you predict your shadow will again be as long as it is at Shadow Position #1?
A. 10
B. 9
C. 8
D. 7

_D_ 10. What do you predict will happen to the length of your shadows as the sun goes from the morning into the afternoon?
A. They will get longer at first and then shorter
B. They will shorter at first and then longer
C. They will get shorter each half hour
D. They will get longer each half hour

_A_ 11. What do you predict will happen to the positions of your shadows as the sun goes from the morning into the afternoon?
A. They will move into positions #5, then #6, then #7, and then #8
B. They will move into positions #9, then #10, then #11, and then #12
C. They will move into positions #12, then #11, then #10, and then #9
D. They will move into positions #8, then #7, then #6, and then #5

GO ON TO THE NEXT PAGE!

Use the following diagram to answer the next 3 questions. The diagram is a graph showing the time candles will burn in jars of differing sizes or volumes.

CANDLES AND JARS: BURN TIME vs VOLUME

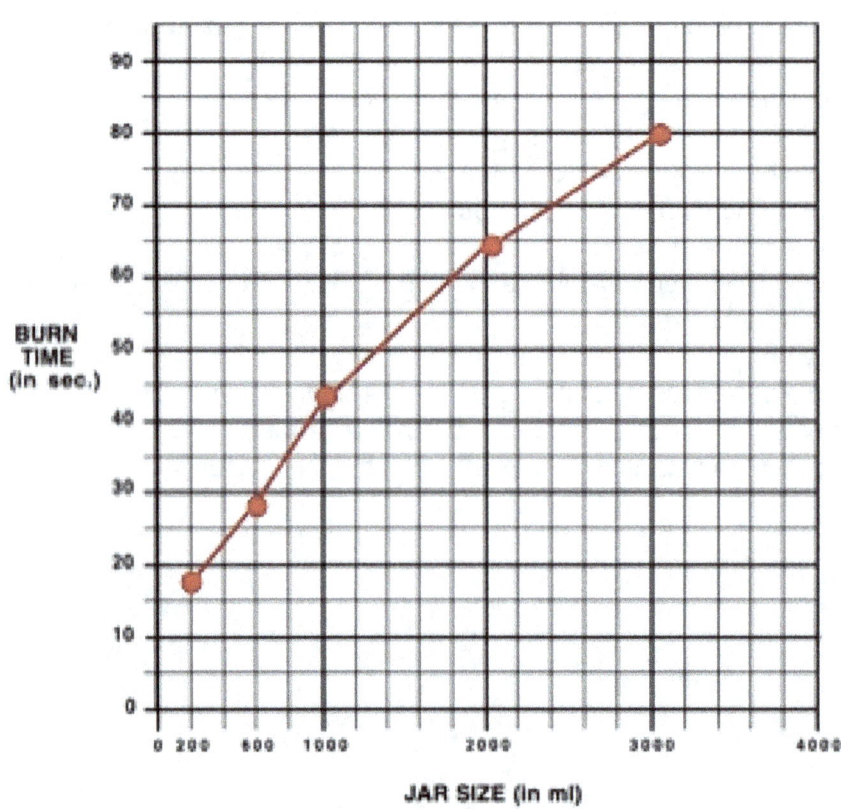

____A____ 12. What do you predict the burn time of a candle will be in a jar that has a volume of 800 ml?
        A. 35 seconds
        B. 38 seconds
        C. 40 seconds

____D____ 13. What do you predict will be the burn time of a candle in a 4,000 ml jar?
        A. it cannot be determined
        B. 80 seconds
        C. 95 seconds
        D. 90 seconds

____A____ 14. What do you predict the volume of a jar would be if a candle's burn time in it is 60 seconds?
        A. 1800 ml
        B. 1500 ml
        C. 1000 ml
        D. 2000 ml

MODULE 14 TEST

THE SCIENCE PROCESS SKILL OF EXPERIMENTING

Module 14 TEST: EXPERIMENTING

NAME: _____

DIRECTIONS: Read each question carefully. Read each answer carefully. Select the best answer and write its CAPITAL letter in the blank to the left of the question. Each is worth 1 point unless otherwise noted. Total test value is 11 points.

_____ 1. Two students were experimenting with how long three candles could stay burning inside different sized jars. Each student took a turn timing the burning of the candles when the jar was place over them. The two students repeated this three times, and each time they came up with different burn times. The problem the students were having with their experiment might have been they didn't . . .

    A. use jars of the proper sizes for the experiment.
    B. create clear and concise predictions about "burn time" for the candles.
    C. operationally define "burn time" for the candles.
    D. establish an experimental control

_____ 2. A group of students was using a ramp and a slider to experiment with how different sized steel balls would roll down the ramp and move the slider. The ramp length was 100 cm and its elevated end was at a height of 5 cm. When each ball was released and rolled down the ramp, it hit the slider and moved it. One student measured the distance the slider moved and recorded it under the inches column in their data table. The problem the students were having with this experiment was that . . .

    A. they needed to run three trials for each ball.
    B. the measuring units did not match.
    C. they forgot to time the slider's movement.
    D. there needed to be better coordination between students doing their jobs.

_____ 3. Three students were building air-powered rockets to see how fin shape might affect the height the rockets could fly. They built one rocket with three triangular-shaped fins, another with four semicircular-shaped fins, and a third with four rectangular-shaped fins. The problem the students were having with this experiment was that they . . .

    A. were not controlling the variables.
    B. forgot to formulate any hypotheses.
    C. predict the rockets' flight height in advance.
    D. failed to read relevant background information about rockets.

GO ON TO THE NEXT PAGE!

_____ 4. Two students decided to study the whitening power of different brands of toothpaste. They got five different brands to use in their experiment. They stained unglazed porcelain squares with beet juice. The students wrote down observations about the toothpastes and then wrote a hypothesis about them. Their hypothesis was that toothpaste #1 would remove the most stain. The problem with these students' experiment was that they . . .

    A. forgot to prepare a data table for their data.
    B. failed to research background information about toothpastes.
    C. picked a poor study to use as an experiment.
    D. wrote a prediction to serve as a hypothesis.

_____ 5. A student decided to study the evaporation rate of water from different kinds of paper towels. The student formulated a problem statement, and then got four kinds of paper towels. The student made observations about each paper towel and decided what equipment was needed to do the experiment. The student was careful to write down a hypothesis about the paper towels and then made some predictions based on that hypothesis. Next, the student moistened the paper towels and set them aside to dry. The next morning the student found all the paper towels were dry. The problem with this student's experiment was . . .

    A. some procedural steps were left out of the plan.
    B. the predictions must not have matched well with the hypothesis.
    C. different paper towels would have produced different results.
    D. the observations about the paper towels were faulty.

_____ 6. Two students liked ants and decided to do an experiment with them. They found an anthill and spent time watching the ants and writing down their observations of the ants' behaviors. The problem the two students had doing this as an experiment was that they . . .

    A. had no way to measure the size of the ant colony.
    B. did not account for other animals in the area where the ants were located.
    C. did not come up with a statement of the problem to investigate.
    D. interfered with the natural cycle of ant life.

GO ON TO THE NEXT PAGE!

_____ 7. A student volunteered to do an experiment on the growth of the two classroom hamsters. One hamster was to be fed a high-protein diet and the other just regular hamster food. Both hamsters were to be weighed every day for a duration of three weeks. The student set up the experiment and each day played a little with each hamster. The student made some notes about how playful each hamster was. The problem with the way the experiment was done was the student . . .

    A. should have checked on the hamsters more than once per day.
    B. needed a partner to share the workload for the experiment.
    C. was biased at the beginning and couldn't do a fair test.
    D. failed to collect and record data on the hamsters but played with them.

_____ 8. A student wanted to do an experiment to find out how the color of paper coverings would affect the heat they collected from sunlight. The student got a white sheet of paper, a pink sheet of paper, a yellow sheet of paper, and a thermometer. Then the student found a sunny spot outside where the papers could be placed. After 20 minutes, the student checked the temperature under each sheet of paper. Then the student realized some other papers should have been used, such as green paper, blue paper, and black paper. So, the student got those papers to add to the outside space. After another 20 minutes, the student went out to check the temperatures again, and realized each piece of paper needed its own thermometer. The problem the student was having with the experiment was . . .

    A. relying on the sun to heat the different colored papers.
    B. not planning for enough or proper materials to use.
    C. selecting 20-minute time intervals to collect data.
    D. not having a partner to help take the temperature readings.

_____ 9. Three students decided to do an experiment to find out how different colors of light would affect the growth of some plants. They made sure to get the same kind and size of plants, and the same size and kind of pots. They prepared a schedule for watering the plants. The students got four lamps and covered one with a blue filter, one with a red filter, one with a yellow filter, and the fourth one with a green filter. They turned on the lamps so each one would shine on just one plant. Then they left the plants alone for several days. After that, the students measured the heights of the plants. The problem with the students' experiment was that they . . .

    A. forgot to establish an experimental control.
    B. should have measured the plants' growth more often.
    C. were not making conclusions based on the plant growth.
    D. should have known plants don't grow well in yellow light.

GO ON TO THE NEXT PAGE!

_____ 10. A student asked to be involved with an experiment on invasive mustard in the meadow behind the school. The student spoke with the scientist leading the study and learned about the ways to estimate invasive mustard plant populations. When the student first went into the meadow, he could not recognize which plants were invasive mustard. The student's problem in this experiment was that he . . .

A. volunteered rather than waiting to be assigned the task.
B. messed up the way to estimate invasive mustard seed plants in the area.
C. had not gained background information about the plant to recognize it.
D. should have done the work in an area farther away from the school.

_____ 11. A student set up an experiment to determine whether or not the temperature of water affected the dissolving rate of a powdered drink mix. The student researched information about dissolving, wrote a statement of the problem, identified the variables, formulated a hypothesis, made some predictions, determined what materials were needed and how to measure the dissolving, outlined the procedural steps to take, and then conducted the experiment. The student carefully recorded all the data in a data table and then graphed the data. Afterward, the student turned in the report to the teacher. The thing the student forgot to include in the report was . . .

A. photographs of the different temperature set-ups.
B. an analysis of the results.
C. acknowledgement of the help other people provided.
D. a title page and index.

Module 14 TEST: EXPERIMENTING

NAME: _____KEY_____

DIRECTIONS: Read each question carefully. Read each answer carefully. Select the best answer and write its CAPITAL letter in the blank to the left of the question. Each is worth 1 point unless otherwise noted. Total test value is 11 points.

C   1. Two students were experimenting with how long three candles could stay burning inside different sized jars. Each student took a turn timing the burning of the candles when the jar was place over them. The two students repeated this three times, and each time they came up with different burn times. The problem the students were having with their experiment might have been they didn't . . .

     A. use jars of the proper sizes for the experiment.
     B. create clear and concise predictions about "burn time" for the candles.
     C. operationally define "burn time" for the candles.
     D. establish an experimental control

B   2. A group of students was using a ramp and a slider to experiment with how different sized steel balls would roll down the ramp and move the slider. The ramp length was 100 cm and its elevated end was at a height of 5 cm. When each ball was released and rolled down the ramp, it hit the slider and moved it. One student measured the distance the slider moved and recorded it under the inches column in their data table. The problem the students were having with this experiment was that . . .

     A. they needed to run three trials for each ball.
     B. the measuring units did not match.
     C. they forgot to time the slider's movement.
     D. there needed to be better coordination between students doing their jobs.

A   3. Three students were building air-powered rockets to see how fin shape might affect the height the rockets could fly. They built one rocket with three triangular-shaped fins, another with four semicircular-shaped fins, and a third with four rectangular-shaped fins. The problem the students were having with this experiment was that they . . .

     A. were not controlling the variables.
     B. forgot to formulate any hypotheses.
     C. predict the rockets' flight height in advance.
     D. failed to read relevant background information about rockets.

GO ON TO THE NEXT PAGE!

_D_   4. Two students decided to study the whitening power of different brands of toothpaste. They got five different brands to use in their experiment. They stained unglazed porcelain squares with beet juice. The students wrote down observations about the toothpastes and then wrote a hypothesis about them. Their hypothesis was that toothpaste #1 would remove the most stain. The problem with these students' experiment was that they . . .

     A. forgot to prepare a data table for their data.
     B. failed to research background information about toothpastes.
     C. picked a poor study to use as an experiment.
     D. wrote a prediction to serve as a hypothesis.

_A_   5. A student decided to study the evaporation rate of water from different kinds of paper towels. The student formulated a problem statement, and then got four kinds of paper towels. The student made observations about each paper towel and decided what equipment was needed to do the experiment. The student was careful to write down a hypothesis about the paper towels and then made some predictions based on that hypothesis. Next, the student moistened the paper towels and set them aside to dry. The next morning the student found all the paper towels were dry. The problem with this student's experiment was . . .

     A. some procedural steps were left out of the plan.
     B. the predictions must not have matched well with the hypothesis.
     C. different paper towels would have produced different results.
     D. the observations about the paper towels were faulty.

_C_   6. Two students liked ants and decided to do an experiment with them. They found an anthill and spent time watching the ants and writing down their observations of the ants' behaviors. The problem the two students had doing this as an experiment was that they . . .

     A. had no way to measure the size of the ant colony.
     B. did not account for other animals in the area where the ants were located.
     C. did not come up with a statement of the problem to investigate.
     D. interfered with the natural cycle of ant life.

GO ON TO THE NEXT PAGE!

D   7. A student volunteered to do an experiment on the growth of the two classroom hamsters. One hamster was to be fed a high-protein diet and the other just regular hamster food. Both hamsters were to be weighed every day for a duration of three weeks. The student set up the experiment and each day played a little with each hamster.   The student made some notes about how playful each hamster was. The problem with the way the experiment was done was the student . . .

A. should have checked on the hamsters more than once per day.
B. needed a partner to share the workload for the experiment.
C. was biased at the beginning and couldn't do a fair test.
D. failed to collect and record data on the hamsters but played with them.

B   8. A student wanted to do an experiment to find out how the color of paper coverings would affect the heat they collected from sunlight. The student got a white sheet of paper, a pink sheet of paper, a yellow sheet of paper, and a thermometer. Then the student found a sunny spot outside where the papers could be placed. After 20 minutes, the student checked the temperature under each sheet of paper. Then the student realized some other papers should have been used, such as green paper, blue paper, and black paper. So, the student got those papers to add to the outside space. After another 20 minutes, the student went out to check the temperatures again, and realized each piece of paper needed its own thermometer. The problem the student was having with the experiment was . . .

A. relying on the sun to heat the different colored papers.
B. not planning for enough or proper materials to use.
C. selecting 20-minute time intervals to collect data.
D. not having a partner to help take the temperature readings.

A   9. Three students decided to do an experiment to find out how different colors of light would affect the growth of some plants. They made sure to get the same kind and size of plants, and the same size and kind of pots. They prepared a schedule for watering the plants. The students got four lamps and covered one with a blue filter, one with a red filter, one with a yellow filter, and the fourth one with a green filter. They turned on the lamps so each one would shine on just one plant. Then they left the plants alone for several days. After that, the students measured the heights of the plants. The problem with the students' experiment was that they . . .

A. forgot to establish an experimental control.
B. should have measured the plants' growth more often.
C. were not making conclusions based on the plant growth.
D. should have known plants don't grow well in yellow light.

GO ON TO THE NEXT PAGE!

C   10. A student asked to be involved with an experiment on invasive mustard in the meadow behind the school. The student spoke with the scientist leading the study and learned about the ways to estimate invasive mustard plant populations. When the student first went into the meadow, he could not recognize which plants were invasive mustard. The student's problem in this experiment was that he . . .

A. volunteered rather than waiting to be assigned the task.
B. messed up the way to estimate invasive mustard seed plants in the area.
C. had not gained background information about the plant to recognize it.
D. should have done the work in an area farther away from the school.

B   11. A student set up an experiment to determine whether or not the temperature of water affected the dissolving rate of a powdered drink mix. The student researched information about dissolving, wrote a statement of the problem, identified the variables, formulated a hypothesis, made some predictions, determined what materials were needed and how to measure the dissolving, outlined the procedural steps to take, and then conducted the experiment. The student carefully recorded all the data in a data table and then graphed the data. Afterward, the student turned in the report to the teacher. The thing the student forgot to include in the report was . . .

A. photographs of the different temperature set-ups.
B. an analysis of the results.
C. acknowledgement of the help other people provided.
D. a title page and index.

www.ingramcontent.com/pod-product-compliance
Lightning Source LLC
Chambersburg PA
CBHW082134290526
45794CB00008B/3026